Springer Proceedings in Materials

Volume 12

Springer Proceedings in Materials publishes the latest research in Materials Science and Engineering presented at high standard academic conferences and scientific meetings. It provides a platform for researchers, professionals and students to present their scientific findings and stay up-to-date with the development in Materials Science and Engineering. The scope is multidisciplinary and ranges from fundamental to applied research, including, but not limited to:

- Structural Materials
- Metallic Materials
- Magnetic, Optical and Electronic Materials
- Ceramics, Glass, Composites, Natural Materials
- Biomaterials
- Nanotechnology
- Characterization and Evaluation of Materials
- Energy Materials
- Materials Processing

To submit a proposal or request further information, please contact one of our Springer Publishing Editors according to your affiliation:

European countries: **Mayra Castro** (mayra.castro@springer.com)
India, South Asia and Middle East: **Priya Vyas** (priya.vyas@springer.com)
South Korea: **Smith Chae** (smith.chae@springer.com)
Southeast Asia, Australia and New Zealand: **Ramesh Nath Premnath** (ramesh.premnath@springernature.com)
The Americas: **Michael Luby** (michael.luby@springer.com)
China and all the other countries or regions: **Mengchu Huang** (mengchu.huang@springer.com)

More information about this series at https://link.springer.com/bookseries/16157

Ranjeet Prasad · Rina Sahu · K. L. Sahoo ·
G. N. Jadhav
Editors

Advancement in Materials Processing Technology

Select Proceedings of AMPT 2020

Springer

Editors
Ranjeet Prasad
NIT Jamshedpur
Jamshedpur, India

Rina Sahu
NIT Jamshedpur
Jamshedpur, India

K. L. Sahoo
CSIR-NML Jamshedpur
Jamshedpur, India

G. N. Jadhav
Department of Earth Sciences
Indian Institute of Technology Bombay
Mumbai, Maharashtra, India

ISSN 2662-3161 ISSN 2662-317X (electronic)
Springer Proceedings in Materials
ISBN 978-981-16-3296-9 ISBN 978-981-16-3297-6 (eBook)
https://doi.org/10.1007/978-981-16-3297-6

This Springer imprint is published by the registered company Springer Nature Singapore Pte Ltd.
The registered company address is: 152 Beach Road, #21-01/04 Gateway East, Singapore 189721, Singapore

Contents

About the Editors

Dr. Ranjeet Prasad is an associate professor and ex-head of Metallurgical and Materials Engineering Department, National Institute of Technology Jamshedpur. He has a Ph.D. in Geology from Ranchi University. His area of expertise is mineral processing technology. He has a teaching experience of 33 years. He has more than 50 publications in national and international conferences. He was the member of BOS, Binova Bhave University, 2014–2017. He is the member of Departmental Research Committee (DRC), 2017–2020.

Dr. Rina Sahu is an assistant professor of Metallurgical and Materials Engineering Department, National Institute of Technology Jamshedpur. She has completed her B.E. and M.Tech. in Metallurgy from NIT Jamshedpur and Ph.D. in Metallic Glasses-IIEST Shibpur. She has specialization in metallic glasses, nanostructured materials, lean grade ores, mineral beneficiation and hydrometallurgy. She has research experience of six years at NML and teaching experience of more than nine years. She has published 110 proceedings papers in international and national conferences and 34 papers in SCI, Scopus-indexed and others journals.

Dr. K. L. Sahoo is a senior principal scientist in materials engineering division CSIR—NML Jamshedpur. He has done his Ph.D. from IIT Kharagpur. He has 21 years of experience in R&D work at various prestigious institutions. His major fields of work include metallic glasses and nanocrystalline materials, quasicrystals, casting, semi-solid processing and forming process foams (steel and Al) and alloy development. He has published 149 papers in respected journals. Some of his notable achievements include Prof. Shilobhadra Banerjee award-best in house OLP project in the year 2016; National Metallurgist Day (NMD) Award in the year 2013; Raman Research Fellowship Award 2008–09; CSIR Young Scientist Award (2000)—in Engineering Sciences for the year 2000; Nijhawan Award (2001)—for best paper published from NML in 2000; INSA—The Royal Society fellowship Award—to visit University of Oxford for three months (2002); JICA fellowship award—to visit Japan for three months (1999); and first prize in metallography contest 1997 IIT Kharagpur.

Prof. G. N. Jadhav is a professor of the Department of Earth Science in the Indian Institute of Technology Bombay, Mumbai. He has done Ph.D. in Geology. He has more than 28 years of teaching experience. His major fields of work include melt inclusion petrography, ore petrology, mineral exploration and mineral beneficiation. He has published ten papers in respected journals. Some of his notable achievements include overall in charge of "Techconnect-2013-14" of "Techfest-2013-14" of IIT-Bombay; Development of Labs.: 05 and Museum-Gemology; Organization of workshops-2; Chairman Organizing Committee of MPT-2016, TCS, Hinjewadi, Pune, from 5th to 7th Jan 2016; Co-Chairman of ACROFI-VI, held at VMCC, IITB from 25th to 27th November 2016; Co-Editor of Proceedings of The International Seminar on "Mineral Processing Technology-2007"; Co-Editor of Proceedings of The International Seminar on "Mineral Processing Technology-2016 and Co-Editor of ACROFI-VI, Extended Abstract Volume, (25th to 27th November 2016).

Al–Ni Duplex Ion Implantation of Nickel Aluminum Bronze (NAB) to Improve the Corrosion Properties

A. Madhubala and K. Thillairajan

1 Introduction

Bronzes being an important copper alloy are used in a variety of atmospheres such as freshwater, seawater, non-oxidizing acids, and industrial fluids. They have a wide variety of applications due to their excellent physical, chemical, and mechanical properties [1]. Alloy bronzes are also used in petrochemical industries, desalination plants, in power generators, aircraft, automotive and railway engineering applications, and electrical and building industries [2]. Nickel aluminum bronze containing more than 9% aluminum forms the β phase. They undergo eutectoid transformation at 565 °C to form the γ phase. The addition of alloying elements as nickel, iron, tin, silicon, and manganese provide better properties [3–5]. Due to alloy additions, along with alpha (α) phase, retained beta (β) phase, and four forms of kappa (κ) phases, namely κ_{I}, κ_{II}, κ_{III}, and κ_{IV} are observed in the nickel aluminum bronze microstructures [6]. Especially addition of iron and nickel in dual-phase alloys develop more kappa (κ) phases suppressing the embrittling γ phase. The β phase is the martensite phase formed due to the rapid cooling of dual-phase NAB alloy. Tempering is carried out to transform the β phase into beneficial kappa phases [7]. By Galvanic series, copper and its alloys are nobler compared to certain ferrous alloys, and the alloys of aluminum, zinc, magnesium. But when used in combination with a few stainless steel grades, titanium, nickel, zinc, or with other noble metals, protection must be provided to copper and its alloys to avoid corrosion. The internal structure of a NAB alloy attacked during corrosion follows the order as γ, β, and then κ phases [8]. The presence of the γ_2 phase leads to dealuminification [9, 10]. Corrosion can be avoided in suitable ways respective for each material or environment. Processes like friction stir processing, ion plating, thermal spraying, and coatings enhance the

A. Madhubala (✉) · K. Thillairajan
Department of Metallurgical Engineering, PSG College of Technology, Coimbatore, Tamil Nadu, India

R. Prasad et al. (eds.), *Advancement in Materials Processing Technology*, Springer Proceedings in Materials 12,
https://doi.org/10.1007/978-981-16-3297-6_1

corrosion behavior but with some issues like coated layer detachment, development of stresses, and many others [11, 12]. The addition of aluminum beyond 10–11% though increases its mechanical properties, does not have any evident improvements in corrosion behaviors. So aluminum content has to be carefully maintained within the range of 9–10%. The addition of nickel during production increases the corrosion behavior of the material but also increases the material cost. These alloys can be added to the substrate without affecting the bulk by a process called ion implantation. It is a surface modification process that offers improved surface and subsurface characteristics. Ions of predetermined energy are irradiated into the substrate by an accelerated ion beam to be imbed at a predetermined depth [13]. This impact is carried out at specific beam current, beam dosage, and beam angle to produce required changes or improvements in the surface properties. Ion implantation exerts improved properties by altering crystal structure and/or forming new intermetallic phases or compounds. Mild steels, tool steels, stainless steels, co-cemented tungsten carbide, polyesters, and many are subjected to ion beam irradiations with the ion of Ce, Cr, Ni, Yt, Ti, C, N, and O to provide enhanced characteristics [14–16]. In the present work, ions of aluminum and nickel are implanted into the surface of nickel aluminum bronze alloy (C95800) and their surface characteristics are explored.

2 Experimental Procedure

Nickel aluminum bronze alloy (C95800), is produced as per the standard ASTM B148. The dual-phase alloy is temper treated between 500 and 715 °C for about 6 h. In this temperature range, the β phase starts to disintegrate. It forms α and κ phases at various temperature ranges. With the progression of time, the formation of other forms of kappa phases takes place. The chemical composition is determined by the optical emission spectroscopic method. Small size NAB samples (20 mm × 20 mm × 3 mm) are metallographically prepared and etched (ethanol containing 5% ferric chloride) for optical and scanning microstructural analysis. X-ray diffraction analysis is taken to gather the phase and crystallographic information.

The samples for ion implantation are well polished and cleaned with acetone/alcohol before loading them into the accelerator. The samples are then irradiated with the ions of aluminum at 2 MeV and nickel at 1.5 MeV using a 1.7MV tandetron accelerator. The accelerator runs at low temperatures under a vacuum of 3.6×10^{-7} mbar. The implanting angle is maintained at 45° with a beam dosage of 5×10^{16} ions per centimeter square. Surface roughness calculations are achieved by performing Atomic Force Microscopy (AFM) under semi-contact mode. The distribution and depth ranges of the irradiated species are determined using Scattering Range of Ions in Matter (SRIM) software. Grazing angle X-ray diffraction (GXRD) experiments are done at 0.5° to study the structural information of the implanted surface. Vickers microhardness measurements are performed at various loads. Electrochemical measurements like potentiostatic polarization and electrochemical impedance spectroscopy are carried out using GAMRY three-cell electrode equipment. The solution used is 3.5% NaCl neutral solution at 25 °C. The counter electrode is platinum wire mesh and Standard Calomel Electrode (SCE) acts as a reference electrode with an area of exposure of 0.25 cm^2.

3 Results and Discussion

3.1 *Characterization of Unmodified Base NAB*

The composition of NAB alloy determined by optical emission spectroscopy is given in Table 1. The presence of aluminum content above 8% prevents selective grain attack. Nickel content above 4% will reduce the possibility of dealuminification. For maximum corrosion resistance, nickel content must be greater than iron content. Iron content between 3 and 5% also increases the solidification range and acts as a grain refiner. Similarly manganese content less than 2% would be beneficial in reducing the selective phase attack of the NAB alloy.

The optical and scanning electron micrographs of base NAB alloy are shown in Fig. 1. Optical images clearly reveal the presence of α phase (rich in Cu) and retained martensitic β phase. Retained martensitic β is due to the presence of manganese which is a β stabilizer. SEM images show the presence of three κ phases. The κ_{II} spheroidized precipitates are present at the α/β grain boundaries, κ_{IV} precipitates are observed inside α grains and κ_{III} phase is observed in form of a continuous lamellar network. Nickel promotes the lamellar growth of κ_{III} which otherwise grows continuously in α grain boundary reducing the corrosion resistance of the alloy. The κ_I phase is not

Table 1 Chemical composition of C95800

Alloying element	Cu	Al	Ni	Fe	Mn	Si	Pb	Co
%	80.740	9.339	4.699	4.349	0.954	0.012	0.001	0.007

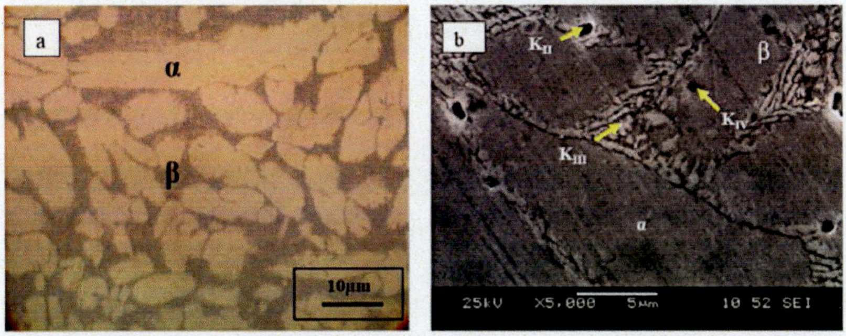

Fig. 1 **a** Optical and **b** SEM images of base NAB alloy

observed in the microstructures as Ni/Fe ratio is > 1 kept with respect to maintaining its toughness and ductility [15].

The energy dispersive spectroscopy (EDS) analysis of base NAB is shown in Fig. 2 and Table 2. It reveals that all the κ phases are a form of nickel-aluminum-iron. Specifically, β phase is Al-rich, κ_{II} is Ni–Fe-rich phase, κ_{III} is Ni–Al-rich phase and κ_{IV} is Ni-rich phase [16]. Ni and Fe present in the κ-phase bind the Al and limit the deleterious β phase formation (Fig. 2).

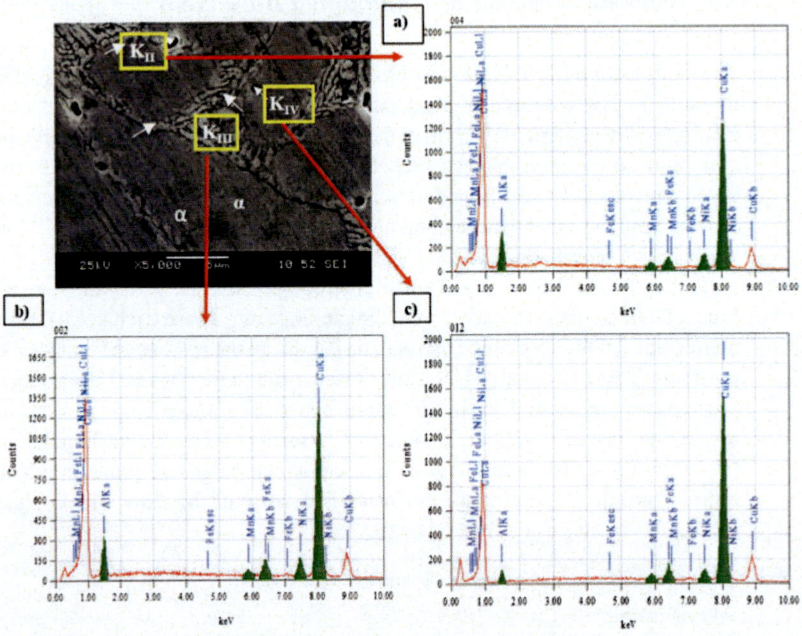

Fig. 2 Elemental spectra for Kappa phases **a** κ_{II}, **b** κ_{III}, and **c** κ_{IV}

Table 2 Mass percentage of major elements in phases present in NAB

Elements\phases	β	κ_{II}	κ_{III}	κ_{IV}
Al	10.35	2.12	5.48	2.21
Ni	3.53	3.87	7.25	4.02
Fe	3.62	3.08	3.53	3.03
Cu	81.9	90.08	82.8	89.8

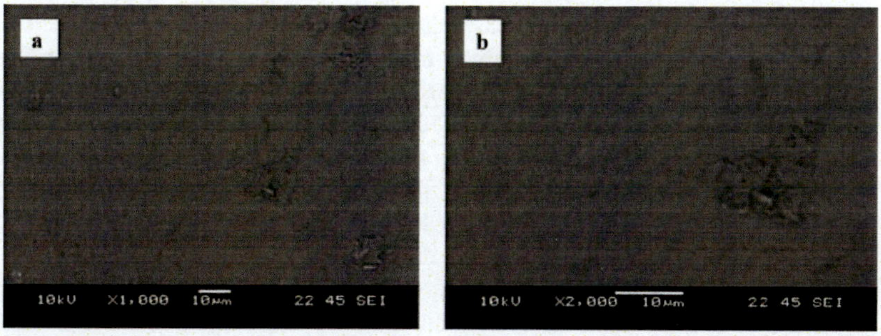

Fig. 3 SEM microscopic images of implanted NAB samples

3.2 *Characterization of Implanted NAB*

The SEM images of the implanted sample shown in Fig. 3 reveal the presence of a few small blisters. These are the effects of high-energy irradiations. The blisters act as low-energy areas initiating corrosion.

Atomic force microscopy (AFM) images of the implanted sample are shown in Fig. 4. Roughness parameters determined from the AFM image are given in Table 3. Roughness modifications of the specimen depend on implantation parameters like ion beam dosage and beam energy. These high-energy irradiations strike the micrometric rough surface and level them. The surface roughness of the implanted specimen is 4 times lowered compared to the unmodified specimen [17]. The reduction in roughness

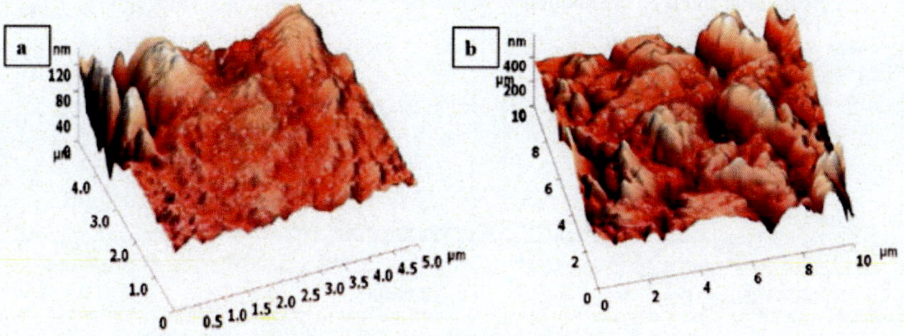

Fig. 4 Surface analysis of implanted NAB samples by AFM

Table 3 Roughness parameters determined from AFM images

Sample	Mean roughness (nm)	RMS roughness (nm)	Skewness	Kurtosis
Unmodified NAB	194.11	275.40	−0.9667	0.4921
Al and Ni implanted NAB	48.48	65.24	−0.3643	11.17789

in the implanted sample increases the energy required for corrosive species to enter the substrate. Theoretical analysis for distribution of irradiated ions by Scattering Range of Ions in Matter (SRIM) software is shown in Fig. 5. It shows that the implanted species has a Gaussian-like distribution with the combined distribution depth range up to 400 nm (200 nm for Ni + 400 nm for Al). The skewness values from AFM interpreted image and SRIM calculations show a negative distribution having its tail extended towards the left. The kurtosis on both infers the high degree of peakedness showing leptokurtic distribution [18].

X-ray diffraction analysis of the NAB samples is shown in Fig. 6. The implanted sample peak positions are quite similar to the peaks of the base sample. It points out that implanted ions are in solid solution with the matrix [4, 5]. However, broadening of

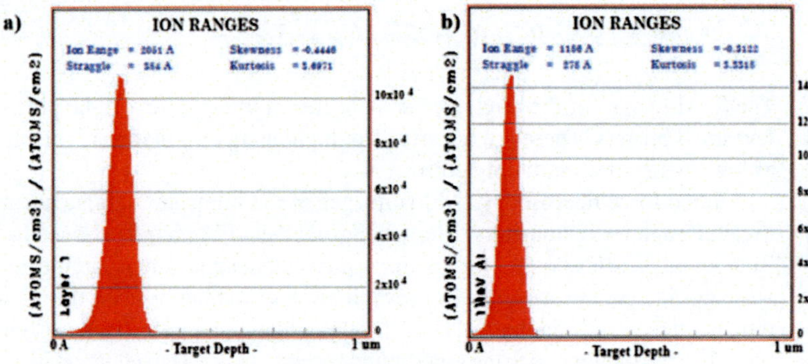

Fig. 5 Depth and Distrbution range of irradiated species in NAB sample by SRIM software **a** Al distribution and **b** Ni distribution

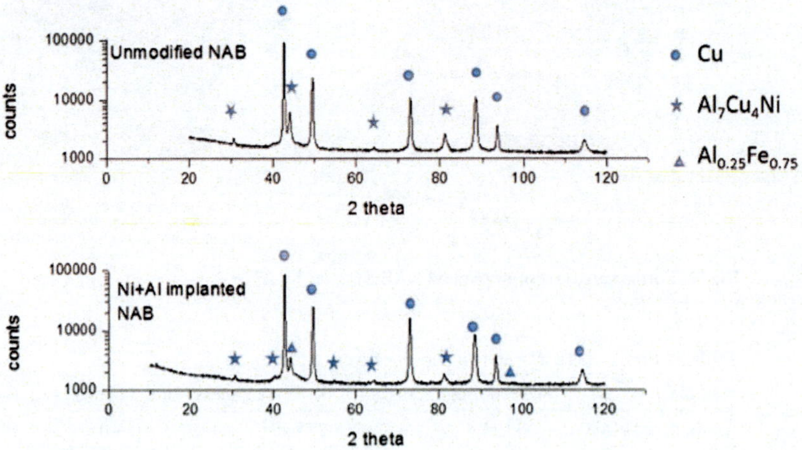

Fig. 6 XRD peaks of unmodified and implanted samples

a few Cu peaks (49.787°, 88.579°, and 114.891°) are observed. The new intermetallic peaks are observed by the formation of Al_7Cu_4Ni and $Al_{0.25}Fe_{0.75}$ compounds. The amplitudes of few peaks (30.885°, 44.171°, 64.243°, 73.083°, 81.221°, and 93.649°) are increased. Peak broadening and increase in amplitudes are mainly due to the ion straggling, which happens due to the cascade collisions that an ion experience at the atomic level. These modifications cause changes in crystallite size (D). Dislocation density and microstrain also increase due to the formation of intermetallics. The decrease in crystallite size also confirms the inverse relation of crystallite size with dislocation density and microstrain [19]. The crystallite size is theoretically calculated using Debye Scherrer equations.

$$\mathrm{D} = \mathrm{K}\,\lambda/\beta\cos\theta \tag{1}$$

where K—shape factor; λ—wavelength, β—full-width half maxima, Θ—peak position.

$$\delta = 1/(\mathrm{D})^2 \tag{2}$$

$$\text{and } \varepsilon = \beta/4\tan\theta \tag{3}$$

The average grain size of the implanted sample is 0.177 nm and that of the unmodified sample is 0.825 nm [20]. The crystallinity of the implanted sample is 46.273%, which is reduced from 62.315% of the unmodified NAB. The average dislocation density increased 1.36 times and the average microstrain increased 0.639 times than the unmodified sample. The Vickers microhardness measurements are carried out on both the NAB samples and are shown in Fig. 7. The implanted specimen has a high surface hardness. An increase in hardness is attributed due to the effect of an increase in dislocation densities by the intermetallics. The increase in hardness is limited to the implanted ion depth [21].

Fig. 7 Hardness versus penetration depth of unmodified and ion implanted samples

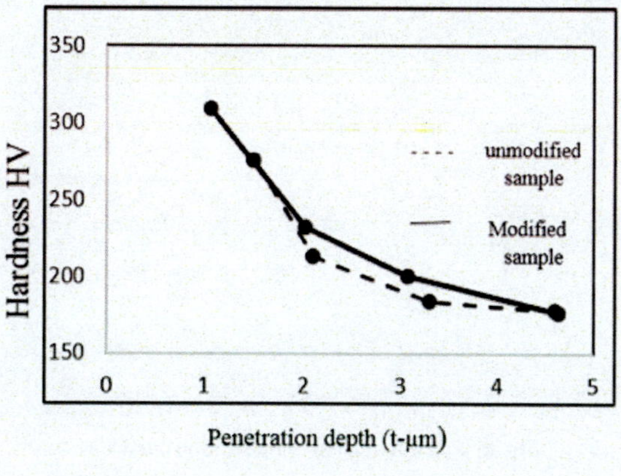

The potentiodynamic polarization corrosion test graphs are shown in Fig. 8. Electrochemical polarization measurements from the polarization curves are determined using Gamry Echem Analyst software and are shown in Table 4. The lower open circuit potential of the samples indicates their weak tendency to corrode. The implanted sample has higher corrosion resistance. The corrosion process in the implanted process is under passivation control and is apparent from the higher anodic slope. The corrosion potential of the implanted sample is increased and its corrosion current has reduced indicating the increased corrosion resistance of the implanted sample. The implanted sample has shown 10 times lower corrosion rate compared to the base alloy. The implanted nickel and aluminum ions have formed impervious and insoluble passive film [11, 12]. The layer of nickel and aluminum implanted also hinder the aggressive corrosive specious from penetrating the bulk to form oxides. The corrosion current required to maintain the passive film is small. The repassivation of the film is rather spontaneous due to the implanted aluminum ions. Hence, the polarization resistance of the modified alloy is very much higher than the base NAB alloy.

Electrochemical impedance plots are shown in Fig. 9a and b. The Bode plot shows that the frequency increases impedance decreases. When the material begins to corrode to form a passive layer, the phase angle drops eventually. On further increase in frequency, the phase angle and impedance would increase till the material becomes passive to the corrosive reactions. This determines the decrease in charge transfer and solution resistance of the alloy upon an increase in frequency. The Nyquist plot indicates the total impedance of the implanted sample which is higher than the unmodified alloy. The left end of the curve tells about the electrical double layer which

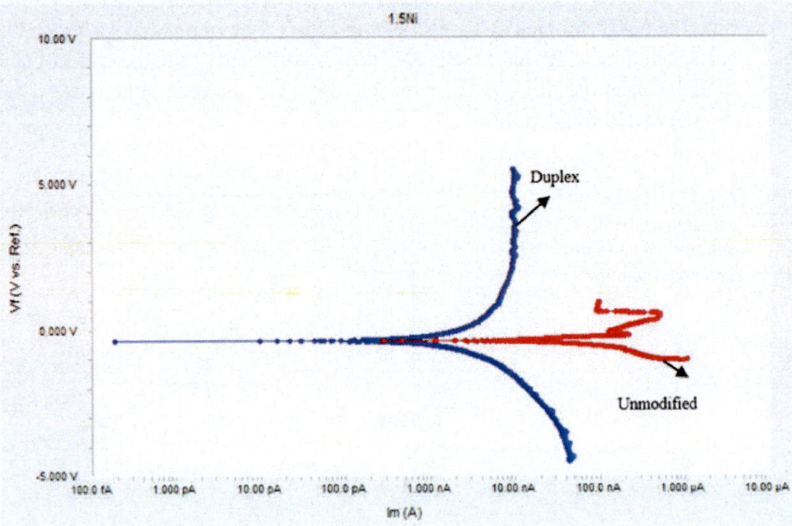

Fig. 8 Potentiodynamic polarization curves

Table 4 Electrochemical polarization measurements of modified and unmodified samples using Gamry Echem analyst software

Parameter	Unmodified sample	Al- and Ni-implanted sample
Open circuit	0.0123659 V	0.612397 V
βa	1.320 V/decade	1e15 V/decade
βc	0.917 V/decade	5.224 V/decade
Icorr	128 nA	12.4 nA
Ecorr	−385 mV versus SCE	−381 mV versus SCE
Rp	1.836 M Ω	152 M Ω
Corrosion rate	228.2 × 10^{-3} mpy (0.006 mm)	22.05 × 10^{-3} mpy (0.0005 mm)

a

b

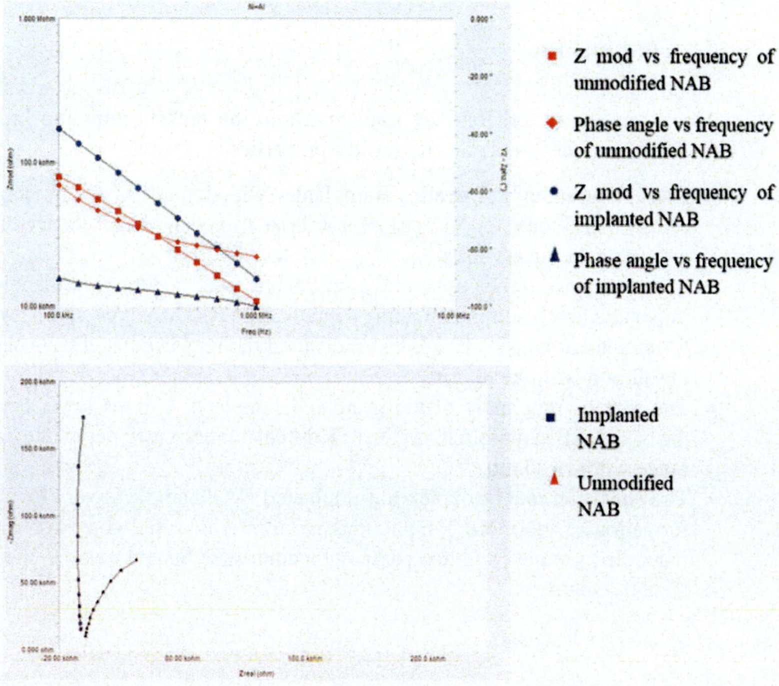

Fig. 9 Electrochemical impedance spectroscopy **a** Bode plot **b** Nyquist plot

acts as a capacitor when the current begins to pass into the substrate. The barrier to the corrosive species increases as the charge transfer and solution resistance increases and thus the material sweeps a bigger curve. The major electrochemical reactions in the passive film formation are the formation of cuprous oxide and copper chloride [22]. The formation of other copper oxides as hydroxides and copper hydroxychlorides takes place on long exposure to corroding media [23]. The presence of the nickel layer compacts the passive film thereby increasing the corrosion resistance. The passive film gets stabilized and impervious due to the presence of implanted aluminum layer which also forms its oxides [24]. Cathodic and anodic reactions involved are

$$O_2 + 4H^+ + 4e^- \rightarrow 2H_2O \tag{4}$$

$$2Cu + H_2O \rightarrow Cu_2O + 2H^+ \tag{5}$$

$$Cu + Cl^- \rightarrow CuCl^- + e^- \tag{6}$$

4 Conclusion

Aluminum and nickel ions are implanted into the nickel aluminum bronze alloy (C95800) surface to enhance its surface properties.

- Nickel aluminum bronze alloy is implanted with duplex (Al and Ni) ions with the energy of 1.5 meV for Ni^+ and 2 meV for Al^+ ions under the fluence of 5×10^{16} ions per centimeters square.
- Microhardness of the implanted sample is increased and the increment is up to the implanted depth without affecting the bulk. The XRD analysis also shows that the formations of new peaks are less indicating the ions irradiated have been mostly in solution with the matrix.
- The surface roughness of the irradiated sample is 4 times lower compared to the base NAB sample. Theoretical SRIM calculations also depict the distribution range within 400 nm.
- The corrosion rate of duplex ion-implanted NAB sample is very low due to the formation of the more compact passive film. Under the corrosive atmosphere, implanted sample exhibited passivation controlled behavior due to the irradiated Ni and Al ions.

References

1. The copper advantage-a guide to working with copper and copper alloys. www.antimicrobialcopper.com
2. Dragoș-Cristian A, Petrică V, Dragoș-Ionuț D, Mirabela-Georgiana M, Viorel G Study of aluminium bronze, mark CuAl9Mn2, TEHNOMUS—new technologies and products in machine manufacturing technologies
3. Meigh HJ (2000) Cast and wrought aluminum bronzes-properties, processes and structure. Institute of Materials, London, p 404

4. Olszewski AM (2008) Dealloying of a nickel-aluminum bronze impeller. J Fail Anal Prev 8:505–508
5. Feest EA, Cook IA (1983) Pre-primary phase formation in solidification of nickel aluminum bronze. Metals Technol 10(1):121–124
6. Schussler A, Exner HE (1993) The corrosion of nickel aluminum bronzes in sea water-1. Prot Layer Form Passiv Mech Corros Sci 34(11):1793–1802
7. Cottam R, Barry T, McDonald D, Li H, Edwards D, Majumdar A, Dominguez J, Wang J, Brandt M (2013) Laser processing of nickel aluminum bronze for improved surface corrosion properties. J. Laser Appl 25:032009. https://doi.org/10.2351/1.4799555
8. Michels HT, Kain RM (2003) Effect of composition and microstructure on the seawater corrosion resistance of nickel aluminum bronze. In: Corrosion, NACE international, paper no. 03262
9. Lenard DR, Bayley CJ, Noren BA (2008) Electrochemical monitoring of selective phase corrosion of nickel aluminum bronze in seawater. In: Corrosion science section, NACE international
10. Culpan EA, Foley AG (1982) The detection of selective phase corrosion in cast nickel aluminum bronze by acoustic emission techniques. J Mater Sci 17:953–964
11. Qin Z, Wu Z, Zen X, Luo Q et al (2016) Improving corrosion resistance of a nickel-aluminum bronze alloy via nickel ion implantation, vol 72. In: Corrosion science section. NACE International; Corrosion, p 10
12. Qin Z, Luo Q, Zhang Q, Wu Z, Liu L, Shen B, Hu W (2018) Improving corrosion resistance of nickel-aluminum bronzes by surface modification with chromium ion implantation. Surf Coat Technol
13. Goorsky M (2012) Ion implantation. ISBN 978-953-51-0634-0
14. Stroosnijder MF (1998) Ion implantation for high temperature corrosion protection. Surf Coat Technol 100(101):196–201
15. Bakkar A, Neubert V (2005) Improving corrosion resistance of magnesium-based alloys by surface modification with hydrogen by electrochemical ion reduction (EIR) and by plasma immersion ion implantation. Corros Sci 47:1211–1225
16. Lloyd DM, Lorimer GW, Ridley N (1980) Characterization of phases in a nickel aluminium bronze. Metals Technol 633C
17. Song QN, Zheng YG, Ni DR, Ma ZY (2014) Study on the nobility of phases using scanning kelvin probe microscopy and its relationship with corrosion behavior in the chloride media for as-cast and friction-stir processed Ni-Al bronze. Corros Sci. https://doi.org/10.1016/j.corsci.2014.11.039
18. Shahnawaz M, Bashir S, Shafique MA, Hussain T (2018) Study the effects of nitrogen ion implantation on structural and mechanical properties of AA7075. Mater Res Express (MRX-107818.R1). https://doi.org/10.1088/2053-1591/aaceb6
19. Natishan PM, McCafferty E (1989) The mechanism of blister formation and rupture in the pitting of ion implanted aluminum. Electrochem Soc Inc. 136:1
20. Wang F, Zheng L, Li Q, Zhang F, Chen X, Zhang H (2018) Corrosion properties of carbon ions implanted chromium coating prepared on CSS-42L aerospace bearing steel. Surf Coat Technol 349:392–99
21. Jayaram P, Pradyumnan PP, Zh. Karazhanov S (2016) Micro-strain, dislocation density and surface chemical state analysis of multication thin films. Phys B Phys Condens Matter (PHYSB309587)
22. Waheed A, Badawy, Rabab M, El-Sherif, Hassan Shehata (2007) Electrochemical behavior of aluminum bronze in sulfate-chloride media. J Appl Electrochem 37:1099–1106
23. Abdul AH, Kadhum AB, Mohamad, (2013) Corrosion of nickel-aluminum-bronze alloy in aerated 0.1 M sodium chloride solutions under hydrodynamic condition. Int J Electrochem Sci 8:4571–4582
24. Yang F, Kang H, Guo E, Li R, Chen Z, Zeng Y, Wang T (2018) The role of nickel in mechanical performance and corrosion behaviour of nickel-aluminium bronze in 3.5 wt.% NaCl solution. Corros Sci 139:333–345

Prediction of Washability Characteristics of Non-coking and Semi-coking Coals

Shaik Saida, Sanchita Chakravarthy, Rina Sahu, Priya Ranjan Mishra, and Koushik Chakravarty

1 Introduction

Non-coking coals are used in the cement, direct reduction, and smelting of iron ore, chemical, paper, etc. industries. India is one of the countries with rich non-coking resources in the world. Most of the Indian coal is formed during natural disasters like floods (drift origin). The major problem with Indian origin is the high ash content [1]. New Zealand coals possess very good petrographical characteristics and low ash content, but the high sulfur content in these coals makes them invulnerable for industrial usage. It is important to determine whether the coal is suitable to wash to lower the ash and sulfur contents with a high percentage of yield [2]. There are so many techniques available for the beneficiation of coals like float and sink, froth flotation, jigging, spiral concentration, cyclones, etc. Among these, float and sink or washability analysis is the primary criteria for evaluating the cleanliness of a coal seam [3, 4]. The data obtained from the sink and float analysis gives a set of curves through which the cleanliness and the cut-off density of respective coal can be obtained easily [5]. Coal cleaning characteristics of coal depend on three main

S. Saida (✉)
Indian Institute of Technology Kharagpur, Kharagpur 721302, West Bengal, India

S. Chakravarthy
Analytical Chemistry Division, National Metallurgical Laboratory Jamshedpur, Jamshedpur 831007, Jharkhand, India
e-mail: sanchita@nmlindia.org

R. Sahu · P. R. Mishra
Metallurgical and Materials Engineering Department, National Institute of Technology Jamshedpur, Jamshedpur 823109, Jharkhand, India
e-mail: reenasahu.met@nitjsr.ac.in

K. Chakravarty
Tata Steel, Jamshedpur 831001, Jharkhand, India

R. Prasad et al. (eds.), *Advancement in Materials Processing Technology*, Springer Proceedings in Materials 12,
https://doi.org/10.1007/978-981-16-3297-6_2

Table 1 Degree of coal washability for coals calculated by using different parameters

No	Coal origin	Parameter	Size	Value (%)	References
1	Ananta OCP (India)	Index of washability (IW)	+25 mm	15.26	Mohanta et al. [6]
			−25 +13 mm	26.56	
			−13 +6 mm	30.65	
			−6 +3 mm	40.81	
			+3 −0.5 mm	42.39	
2	Jamadoba (India)	Washability index (WI)	−3 +0.5 mm	27.12	Sahu et al. [7]
3	Kuju (India)	Washability index (WI)	−75 mm	30.9	Chandrasekhar et al. [8]
			−25 mm	31.9	
			−13 mm	34	
			−6 mm	41.4	
4	Arguvan-lignite (Turkey)	Index of washability (IW)	+0.5 mm	12.85	Aksogan Korkmaz and Bentli [9]
5	KCZ-5 Pakistan (P)	Washability number (W_n)	−9.5 +2.36 mm	46	Nasir et al. [10]
	KCZ-10 (P)		−25 +12.5 mm	22	
	SCZ-8 (P)		−25 +12.5 mm	35	
	KCW-2 (P)		−9.5 +2.36 mm	36	
	KRC-4 (P)		−9.5 +2.36 mm	8	

parameters such as coal liberation, coal washability, and separator performance. Coal consists of organic matter which has a low density of 1.2–1.5 g/cc and mineral matter or shale with a high density of 2.4–2.8 g/cc. Therefore, sink and float analysis is usually carried out in liquid densities of 1.2–2 g/cc.

Previously many researchers have investigated the coal washability characteristics based on the effective liberation of organic matter from the mineral matter content by using qualitative analysis. O'Brien et al. [11] described the washability of coal characteristics based on coal liberation using the fractional size distribution of coal. This technique is qualitative, and liberation characteristics were inferred from the actual data. For decades, microscope techniques are most popularly used for the determination of coal liberation based on the particle size distributions. However, these microscopy techniques are usually giving nearly estimations and cannot be calculated into real figures, and all these microscopy techniques required high equipment costs [12, 13]. Many other researchers have reported on the washability characteristics based on coal liberation. All these techniques involve either qualitative analysis or a lot of graphical representation, which creates problems in terms of accuracy in calculations [14–16]. Looking after the limitation in qualitative analysis, some researchers have worked on the development of simple mathematical equations for representing the coal liberation characteristics from sink and float data. Nasir et al. [10] described the washability characteristics of Pakistan-origin coal in terms of washability number (Wn), which varies between 0 and 100. Govindarajan and Rao [17], Austin [18] developed an index of washability (IW) and locking index (LI) parameters, respectively, for calculating the degree of coal liberation using simple mathematical expressions. In continuation to the IW parameter index, a new parameter, near gravity material index (NGMI), has been proposed by many researchers for calibrating the coal washability based on coal liberation and near gravity material (NGM) [19]. Table 1 describes some of the literature reviews on coal washability characteristics based on different parameters. In the present work, the parameters index of washability (IW) and near gravity material index (NGMI) were utilized to evaluate and compare the washability characteristics of two coals NCC and SCC. The primary reason behind choosing the IW and NGMI parameters for the present data is that both the parameters are in numerical representation for indicating the degree of washing, both the parameters are calculated simultaneously using a single graph. Moreover by using the NGMI the cut-off density for a particular coal can be predicted more accurately.

2 Mathematical Equations

In present calculations, the mathematical equations developed by Govindarajan and Rao [17] and Majumder and Barnwal [19] were used to study the washability characteristics of two coals. They divided the sink and float data into two parts; float material as non-ash (Rna) part and sink material as the ash (Ra) part. Rna and Ra are expressed as follows:

$$R_{na} = \frac{100 \cdot X \cdot (100 - A_i)}{\left(100 - A_f\right)} \tag{1}$$

$$R_a = \frac{100 \cdot X \cdot A_i}{A_f} \tag{2}$$

where X is the cumulative fractional weight of the float at ith relative density, Ai and Af are the cumulative ash content of the float material at ith relative density and the percent ash content of the feed, respectively.

They described that Rna and Ra curves can be fit into the cubic equations given below.

$$R_{na} = \mathrm{x}\mathrm{X} + \mathrm{y}X^2 + zX^3 \tag{3}$$

$$R_a = \mathrm{m}\mathrm{X} + \mathrm{n}X^2 + oX^3 \tag{4}$$

The values of the constants a, b, p, q, x, y, z, m, n, o in the above equations have been estimated for coals 1 and 2 using the least-squares method.

Index of washability (IW), near gravity material (NGMI) of a calculated by using the Rna, Ra curves are very useful full in comparing the coals.

$$\mathrm{IW} = \frac{\{6x + 4y + 3z - 600\}\{6(x - m) + 4(y - n) + 3(z - o)}{3600\ H_A} \quad (5)$$

NGMI of the coal is calculated by the fraction of the area between the ± 0.1 specific gravity range to the total area between the Rna and Ra curves.

$$\mathrm{NGMI} = \frac{Area\ between \pm 0.1\ specific\ gravity\ in\ R_{na}\ and R_a\ curves}{Total\ area\ between\ the\ R_{na}\ and\ R_a\ curves} \quad (6)$$

3 Experimental Work

New Zealand and Indian origin coals collected from seams were put through crushing and grinding operations to liberate the ash and non-ash parts. These ground coals were screened and the coal between −1 and +0.5 mm mesh were collected for the washability study. The washability studies of −1 +0.5 mm size fraction coal were determined by the sink and float method. Sink and float studies were carried out in between the specific densities 1.2 and 2 g/cc. Toluene, benzene, and bromoform were used to prepare the required density fluid medium. This test was carried out by placing the coal samples in a progressively increasing specific density fluid medium and collecting the sink and float fractions from each density fraction. These sink and float fractions were washed, dried, and weighed and before the proximate analysis to determine the ash content.

4 Results and Discussions

Tables 2 and 3 show the sink and float density separation data for NCC and SCC coals, respectively. NGM for the NCC coal is calculated as 12.53, 5.24, 8.07, 17, and 79.4 for specific densities 1.4, 1.5, 1.6, 1.7, and 1.8, respectively. For SCC, it is noted as 62.17, 62.27, 12.22, and 30.7 at specific densities 1.25, 1.3, 1.35, and 1.4, respectively. The coal distribution data from sink and float analysis produces a set of curves (float curve, ash characteristic curve, and NGM curves), which can be used to predict the washability characteristics of coal cleaning. If the float curve has an "L" shape, then the coal is said to be easy washing, and if the NGM at a specific density fraction increases then that coal is said to be difficult to wash at that fraction. Figure 1 shows the variation of NGM with respect to the specific density of NCC and SCC coals. It shows that the slope of the curve varies abruptly at a specific density of 1.7 for NCC coal, for SCC the slope changes at 1.3 and 1.35 specific densities. Figure 2

Table 2 Sink and float analysis of NCC coal

Specific density	Yield weight %	% Ash	Cumulative % of floats	Cumulative float ash %	Cumulative % of sinks	Cumulative sink ash %	NGM
1.4	10.36	6	10.36	6	100	47.85	12.53
1.45	2.17	9.6	12.53	6.65	89.64	52.38	5.24
1.55	3.07	14.6	15.6	8.26	87.74	53.61	8.07
1.65	5	19	20.6	10.84	84.4	54.85	17
1.75	12	27	32.6	16.8	79.4	57.11	79.4
2.2	67.4	62.47	100	47.85	67.4	62.47	–

Table 3 Sink and float analysis of SCC coal

Specific density	Yield weight %	% Ash	Cumulative % of floats	Cumulative float ash %	Cumulative % of sinks	Cumulative sink ash %	NGM
1.225	7.025	0.445	7.025	0.445	100	3	62.17
1.25	55.15	0.76	62.18	0.724	92.85	3.2	62.27
1.3	7.12	3.32	69.3	0.99	36.45	6.74	12.22
1.35	5.1	4.12	74.4	1.21	29.16	7.53	30.7
1.4	25.6	8.21	100	3	23.94	8.21	–

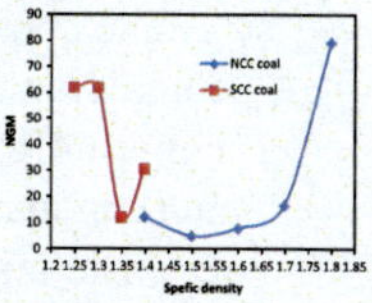

Fig. 1 Specific density versus NGM curves for NCC and SCC coals

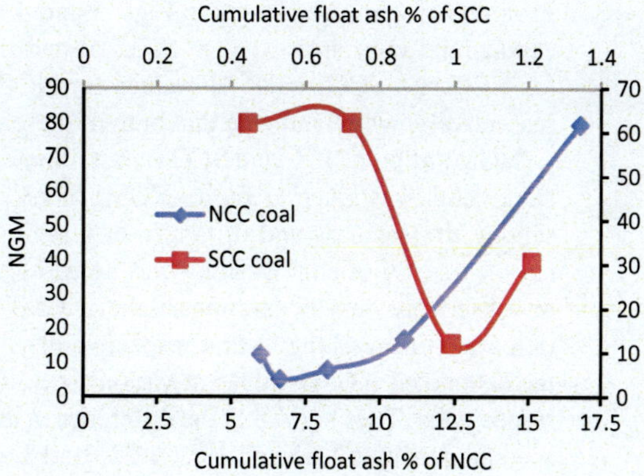

Fig. 2 Cumulative float ash % versus NGM curves for NCC and SCC coals

Table 4 Calculated R_{na} and R_a for NCC data from sink-float data

Avg. specific density	R_{na}	R_a	X
1.4	18.73	1.295	0.104
1.45	22.49	1.74	0.125
1.55	27.52	2.68	0.156
1.65	35.32	4.65	0.206
1.75	52.16	11.41	0.326
2.2	100	100	1

Table 5 Calculated R_{na} and R_a for SCC data from sink-float data

Avg. specific density	R_{na}	R_a	X
1.225	7.21	1.04	0.07
1.25	63.64	15.01	0.62
1.3	70.74	22.87	0.69
1.35	75.77	30.01	0.74
1.4	100	100	1

shows the variation of NGM with respect to the cumulative float ash content for SCC and NCC coals.

The concepts developed by Majumder and Barnwal 2004 [18] were adopted for comparing the washability characteristics of NCC and SCC coals. Variation of the non-ash (Rna) and ash (Ra) fractions of NCC and SCC coals with specific densities and cumulative fractional weight are shown in Tables 4 and 5, respectively. The pre-exponential constants required for calculating the IW and NGMI were calculated by using the curve fitting method for the plots of Rna and Ra against the respective X value, which were plotted in Figs. 3 and 4. Table 5 indicates the calculated pre-exponential constants (x, y, z, m, n, o) and the index of washability (IW) values for NCC and SCC coals. IW values for NCC and SCC are 37.06% and 37.12%, respectively, which indicate that both of these coals were hard to wash.

NGMI of both NCC and SCC were estimated from Figs. 3 and 4 using the expression which was given in Eq. 6. NGMI index of the coals varies from 0 to 1 representing the easiness and difficulty of washing coal, respectively. The appropriate cut-off density of coal washing was noted by the difference between NGMI values of subsequent density fraction, higher NGMI difference indicate the ease of separation in coal washing at that respective density fraction. Tables 6 and 7 represent the calculated NGMI values at various specific densities for NCC and SCC coals, respectively. They show that the difference in the NGMI is higher in between average specific densities 1.65 and 1.75, 1.225 and 1.25 for NCC, SCC coals, respectively. Figure 5 and 6 shows the NGMI variations with respect to average specific density and clean coal ash content for NCC and SCC coals. From the figures, it can be seen that NGMI values are lower at low specific densities and with an increase in the specific densities NGMI raised drastically. It has been seen that both NCC and SCC coals were difficult to wash by density methods. Figure 7 indicates the correlation

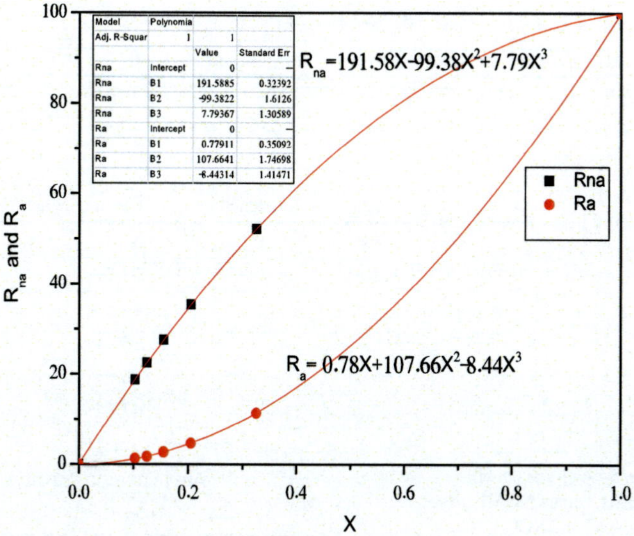

Fig. 3 R_{na} and R_a curves for NCC

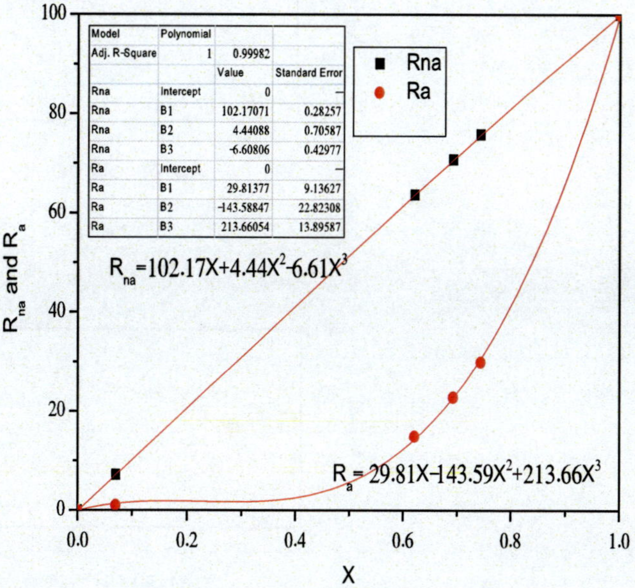

Fig. 4 R_{na} and R_a graphs for SCC coal

Table 6 Coefficients for R_a, R_{na} and Index of washability for NCC and SCC coals

Constants/coal	x	y	z	m	n	o	WI (%)
NCC	191.58	−99.38	7.79	0.78	107.66	−8.44	37.06
SCC	102.17	4.44	−6.6	29.81	−143.59	213.66	37.12

Table 7 Near gravity material index for NCC coals

Avg. specific density	Clean coal ash	NGMI
1.4	6	0.031
1.45	6.65	0.044
1.55	8.26	0.068
1.65	10.84	0.113
1.75	16.8	0.26
2.2	47.85	1

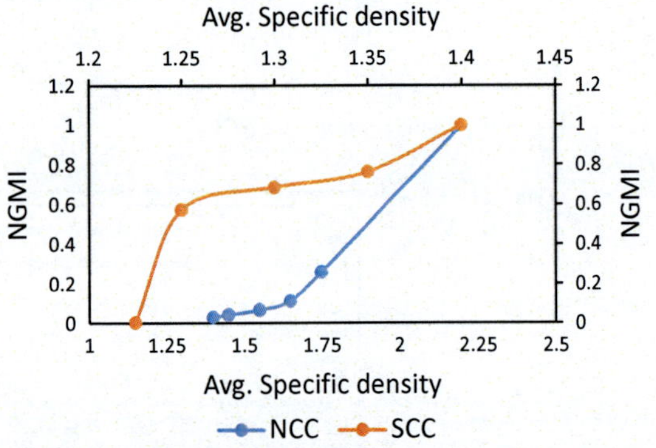

Fig. 5 Average specific density versus NGMI curves for NCC and SCC

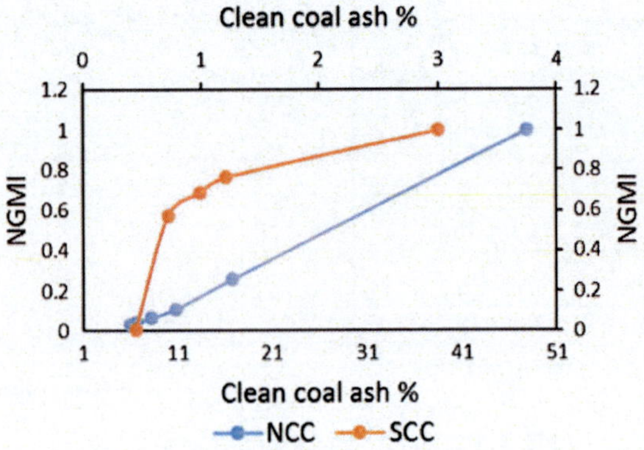

Fig. 6 Clean coal ash % versus NGMI for NCC and SCC

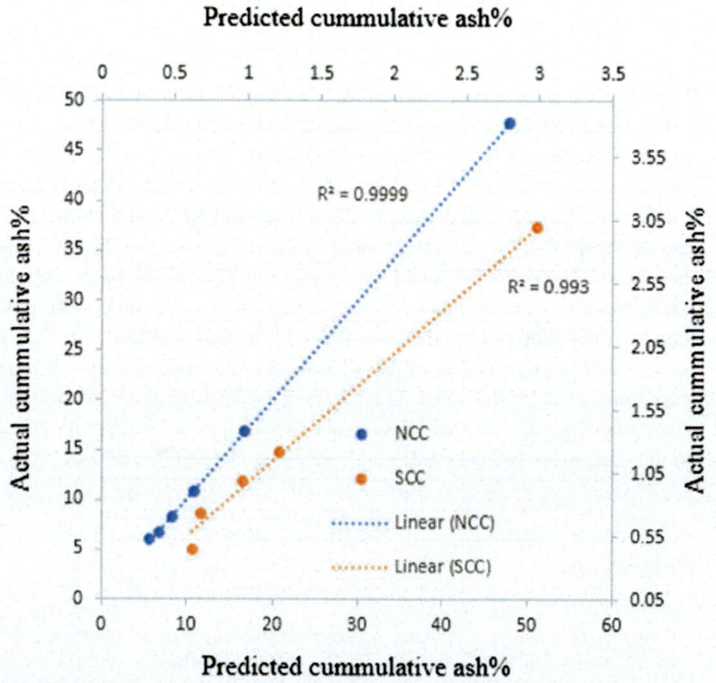

Fig. 7 Comparison of cumulative ash % calculated from experimental results and present mathematical model

between the actual cumulative ash percentage calculated by using the sink and float analysis and the predicted cumulative ash percentage using mathematical expression. The correlation coefficients for NCC and SCC were observed as 0.999 and 0.993, respectively, which indicate that the adopted mathematical expressions were giving good correlation (Table 8).

Table 8 Near gravity material index for SCC coals

Avg. specific density	Clean coal ash	NGMI
1.225	0.445	0.006
1.25	0.724	0.575
1.3	0.99	0.688
1.35	1.21	0.765
1.4	3	1

5 Conclusion

Washability characteristics for the SCC and NCC coals are calculated by using both conventional NGM curves and mathematical equations NGMI curves. It has been observed that through conventional NGM curves the appropriate cut-off density for NCC coal is 1.65 g/cm3. Through mathematical equations, it has been observed that at 1.65 g/cm3 clean coal ash percentage is around 10.84 and NGMI is 0.113, which can be acceptable for density separation for NCC coal. NGM curves for SCC coal show that 1.3 g/cm3 is the cut-off density. For SCC coal, the coefficients calculated by mathematical equation through curve fitting have high standard errors which can be seen in Fig. 4. Through mathematical equations, cut of density for SCC coal is noted as 1.25 with a high NGMI value, which indicates that SCC coal is very difficult to wash compared to NCC coal. The comparative results of cumulative ash percentage calculated using the sink and float analysis and present mathematical model exhibits good correlation, with accuracy over 99% for both NCC and SCC coals.

References

1. Jyoti D, Bhattachrya S, Anupam A, Saxena VK (2015) Washing of low volatile coking (LVC) coal: is that difficult? Trans Indian Inst Met 68(4):649–660
2. Beamish BB (1994) Proximate analysis of New Zealand and Australian coals by thermo-gravimetry. NZ J Geol Geophys 37(4):387–392
3. Longwell JP, Rubin ES, Wilson J (1995) Coal: energy for the future. Prog Energy Combust Sci 21(4):269–360
4. Xu G, Bu X, Mao Y, Ni C, Peng Y, Xie G (2020) Combined column and cell flotation process for improving clean coal quality: laboratory-scale and industry-scale studies. Energy Sources Part A Recover Util Environ Eff 42(21):2678–2687
5. Mir F (2014) Washability characteristics of low volatile Pakistani coking coal by crushing. J Miner Mater Charact Eng 2(05):502
6. Mohanta S, Sahoo B, Behera ID, Pradhan S (2016) Effect of crushing on near-gravity material distribution in different size fractions of an Indian non-coking coal. J South Afr Inst Min Metall 116(2):209–213
7. Sahu D, Chaurasia RC, Suresh N (2019) Mineralogical characterization and washability of Indian coal from Jamadoba. Energy Sources Part A Recover Util Environ Eff 41(5):517–526
8. Chandrasekhar S, Sarkar B, Das A, Chattoraj US, Kumar V, Rao KV, Bhattacharyya KK (2007) Analysis of the washability characteristics of low-volatile Indian coking coal with crushing at different top sizes-a case study
9. Aksogan Korkmaz A, Bentli İ (2017) Determination of washability characteristics of Arguvan-Malatya lignite by different washability index methods. Energy Sources Part A Recover Util Environ Eff 39(14):1572–1580
10. Nasir S, Kucerik J, Mahmood Z (2012) A study on the washability of the Azad Kashmir (Pakistan) coalfield. Fuel Process Technol 99:75–81
11. O'Brien G, Firth B, Adair B (2011) The application of the coal grain analysis method to coal liberation studies. Int J Coal Prep Util 31(2):96–111
12. Shahzad M, Ali Z, Majeed Y, Emad Z, Aaqib M, Adeel B (2016) Liberation studies of Padhrar coal by using fractionation method, XRD analysis and megascopic and microscopic techniques. Pak J Sci Ind Res Ser A Phys Sci 59(2):90–95

13. Ward CR (2016) Analysis, origin and significance of mineral matter in coal: an updated review. Int J Coal Geol 165:1–27
14. Iveson SM, Hunter DM, Galvin KP (2015) A water-based method for measuring density-based partition curves of separators used in coal and mineral processing. Miner Eng 79:196–211
15. Shahzad M, Ali Z (2018) Development of simple techniques for calculating the extent of coal cleaning-part 1: estimating coal liberation characteristics. Int J Coal Prep Util 1–14
16. Cheepurupalli NR, Charan TG, Singh CS, Sharma GVS, Rao ED (2016) Liberation characteristics of indian coking coal through washability investigations. Int Res J Earth Sci 4(1):7–12
17. Govindarajan B, Rao TC (1994) Indexing the washability characteristics of coal. Int J Miner Process 42(3–4):285–293
18. Austin LG (1994) Patterns of liberation of ash from coal. Min Metall Explor 11(3):148–159
19. Majumder AK, Barnwal JP (2004) Development of a new coal washability index. Min Eng 17(1):93–96

Modeling and Simulation of Flashless Forging of Coupling Flange

Mathala Prithvi Raj, Manoj Kumar, and Ajit Kumar Pramanick

1 Introduction

Near-net shape processes have gained a lot of attention in recent years. They are replacing the conventional manufacturing process like casting and forging. The main advantage of the near-net shape process is their ability to produce the final product or a product with the least post-processing, resulting in less wastage of material and reduced manufacturing time. A slight modification to the conventional processes can also result in a near-net-shape process. One such process is flashless forging in which the conventional flash is eliminated by completely enclosing the component in a single die. Flash is the excess material that has to be trimmed out of the component after forging, which is a wastage [1]. In flashless forging, dies are designed in such a way that flash formation is restricted. However, thin fins or burrs may be extruded in the clearance between the dies, which can be removed easily by a hand grinder. So, there is almost no waste of material, and a near-net shape component can be obtained. The most important factor which has to be taken into consideration in the flashless forging is the volume of the billet material, as an excess of it increases the load greatly and less volume results in under-filling.

Several researchers worked on the flashless forging of different components like connecting rod [2, 3], gear blanks [4] and bevel gears [5]. Teruie Takemasu et al. [6], investigated the metal flow of a connecting rod which is produced by flashless forging, and also optimized the preform shape for the production of flashless connecting rods. In all the cases either preform is optimized or a design procedure is provided for the flashless forging of the components. As the problem consists of the optimization of multi-objective responses, several methods like regression analysis or neural networks in conjugation with natural evolution algorithms, response

M. P. Raj · M. Kumar (✉) · A. K. Pramanick
Department of Forge Technology, National Institute of Foundry and Forge Technology, Ranchi, Jharkhand 834003, India

R. Prasad et al. (eds.), *Advancement in Materials Processing Technology*,
Springer Proceedings in Materials 12,
https://doi.org/10.1007/978-981-16-3297-6_3

surface analysis [7, 8] can be used. But using these processes is a laborious process as it requires more experiments. Instead, a modified Taguchi method can be used to optimize the process parameters, where in all the parameters are converted to a comparable grade. The Grey–Taguchi method is used by several types of research [9, 10], where they optimized the response parameters like surface roughness and metal removal rate [9] in a milling operation, the quality parameters optimization in pulsed metal inert gas welding [11], optimization of process parameters in aluminum rib-web forging [12].

In the present work, an attempt is made to design dies for manufacturing of coupling flange by flashless forging. Flanges are used for connecting two pipes, pumps, valves, and in power transmission. They are either welded or bolted. Initially dies are made for conventional forging using the general procedure for design in dies. Three major modifications are done in dies design to convert the dies into flashless dies. Then for both conventional and flashless forgings, an L16 orthogonal array is established. The input parameters considered for conventional forging are billet temperature (°C), die temperature (°C), blocker flash thickness (mm) and friction coefficient, whereas for flashless are billet temperature (°C), die temperature (°C), and friction coefficient. The output or response parameters estimated in both the processes are blocker load (N), blocker effective stress (MPa), blocker wear depth (mm), finisher load (N), finisher effective stress (MPa), blocker wear depth (mm). All the output parameters are converted to grey relation grades (lower the better criteria) and ranked (higher the better) based on the grade. These grey relation grades are converted to S/N ratios considering the higher the better criteria. Finally, analysis of variance (ANOVA) is carried out to find the most significant factor in both processes. Modeling of the dies is carried out in CATIA-V5R19, simulations are carried out in DEFORM-3D, and data analysis is carried out using Minitab 19.

2 Experimental Studies

2.1 Die Design for Conventional Forging

For the designing of forging process, a backward approach has to be used, which means that the last stage has to be designed first and the preceding step is designed based on the shape of the already designed stage. In the present work for forging of coupling flange, three-stage forging is considered. The first stage is an upsetter, the second stage is blocker, and finally finisher. As a backward approach is used initially finisher dies are designed. For modeling of dies, CATIA-V5R19 is used.

Initially, the machining drawing of a coupling flange is converted into a forging drawing by adding all the necessary allowances, draft, fillet, and corner radius according to IS 3469–1 to 3 (1974). Then the finisher drawing is made by adding shrinkage allowance, which is necessary to compensate for the volumetric contraction that takes place while the billet cools from higher temperature to lower temperature, after forging. Based on the finisher drawing, a finisher model is generated and corresponding dies are also generated. Then the blocker drawing is generated based on the finisher model by increasing the internal dimensions and decreasing the external dimensions by 0.3 mm per side and the corresponding dies are modeled. Finally, the upsetting dies are designed. Corresponding models are shown in Fig. 1, as the component is symmetric only 1/4th of the dies are shown. Forging die design is briefly explained in the previously published paper [13] by the authors and the same procedure is followed. Flash thickness and width for finisher are calculated by using Neuberger and Mockel's empirical formula [14] in Eqs. 1 and 2 which gives a flash thickness of 3.6 mm and flash width of 10.8 mm.

$$\text{Flash thicknesst} = 0.89\sqrt{W} - 0.017W + 1.13 \tag{1}$$

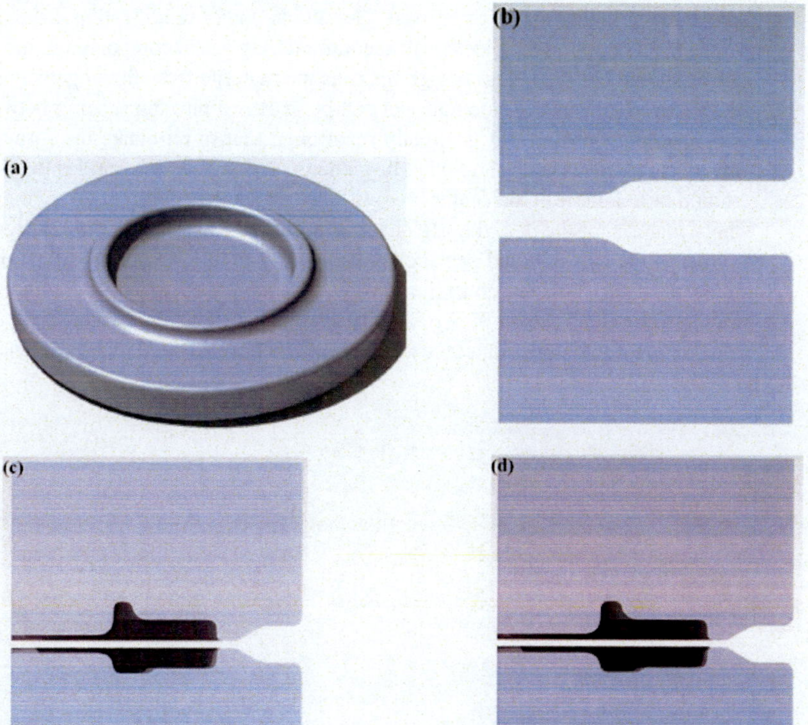

Fig. 1 3D models of **a** Forging **b** Upsetting die set **c** Blocker die set **d** Finisher die set

$$\text{Flash width} w = t\left(3 + 1.2 \times e^{-1.09W}\right) \quad (2)$$

where t = flash thickness (mm), w = flash width (mm), W = Forging weight (kg).

2.2 Die Design for Flashless Forging

In flashless forging, the component is forged without the formation of any flash but thin fins and burrs are formed at the end of the forging which is removed by hand grinder. In the present work, interest is drawn toward flashless forging coupling flange without formation of fins and burrs. For this, some modifications are carried out in the design of the dies compared to that of the conventional forging. The first major modification is the elimination of the upsetting stage. If there is an upsetter then more material will be gathered at the flange area, which may result in the fins/burrs, hence the upsetting stage is removed. The second major modification is the removal of the flash region in the blocker by accommodating entire component in the bottom die as shown in Fig. 2. This design restricts the material from flowing away into the flash region, however, some material can be extruded into the spacing between the dies forming fins/burrs. It is already mentioned that to eliminate the formation of fins/burrs more material should not be gathered at flange which implies that material should be available at the center (peg) till the last stage of forging. To accommodate more material at the peg, the peg shape is modified as shown in Fig. 2a, compared to that of the conventional peg shown in Fig. 1. The third major modification is the removal of the flash region in the finisher by accommodating entire component in the bottom die as shown in Fig. 2b. Other minor modifications which are done in blocker are—to compensate excess material which is provided in the peg, maximum

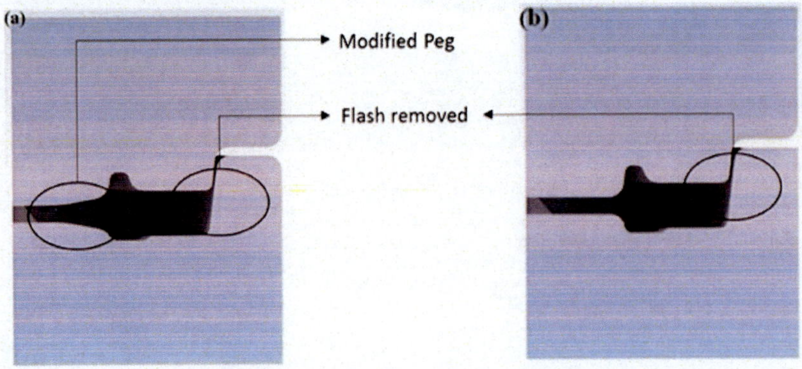

Fig. 2 3D models of **a** Blocker die set **b** Finisher die set

Table 1 Input parameters for conventional forging

Parameters/level	Level 1	Level 2	Level 3	Level 4	Units
Blocker flash thickness	3.2	3.6	4.0	4.4	mm
Friction coefficient	0.3	0.4	0.5	0.7	–
Die temperature	150	200	250	300	°C
Billet temperature	1000	1050	1100	1150	°C

Table 2 Input parameters for flashless forging

Parameters/level	Level 1	Level 2	Level 3	Level 4	Units
Billet temperature	1000	1050	1100	1150	°C
Die temperature	150	200	250	300	°C
Friction coefficient	0.3	0.4	0.5	0.7	–

diameter of the blocker is reduced accordingly and also the depth of the blocker is increase slightly with a corresponding decrease in diameter.

2.3 Materials and Design of Experiments

To estimate the defects like underfilling and folds, commercially available finite element based software DEFORM-3D is used. The flange material is taken as stainless steel AISI-304 with a hot forging temperature range from 900 to 1150 °C. And the die material is taken as H13 tool steel with 50 HRC. The output parameters—load, effective stress, wear depth are estimated for every set of input parameters.

For any component manufactured by forging, it is desirable that the stress induced is uniform all over the component, so, while considering output parameter, effective stress is taken as the difference between the maximum and minimum of von Mises stresses induced in the component after forging. 4 levels of input parameters are considered which are shown in Table 1 and Table 2 for conventional forging and flashless forging, respectively.

L16 orthogonal array is used for the design of experiment for 4 levels and 4 factors in case of conventional forging, and 4 levels and 3 factors in case of flashless forging. Simulations and their corresponding results are shown in Tables 3 and 4 for conventional and flashless forging, respectively.

3 Optimization Using Grey Relation

The number of output parameters that are to be optimized is 6, out of which there are blocker outputs and 3 finisher outputs. A simple Taguchi technique cannot be

Table 3 Design of experiments using L16 orthogonal array for conventional forging

Exp. no	Flash thickness (mm)	Friction coeff.	Die temp. (°C)	Billet temp (°C)	Blocker			Finisher		
					Load (N)	Eff. stress (MPa)	Wear depth (mm)	Load (N)	Eff. stress (MPa)	Wear depth (mm)
1	3.2	0.3	150	1000	16,300,000	300	0.3400	11,900,000	287	0.01240
2	3.2	0.4	200	1050	11,000,000	254	0.0528	11,600,000	249	0.00966
3	3.2	0.5	250	1100	8,970,000	236	0.0482	10,200,000	222	0.00986
4	3.2	0.7	300	1150	9,250,000	221	0.0425	8,800,000	210	0.00930
5	3.6	0.3	200	1100	9,110,000	245	0.0414	9,390,000	235	0.01020
6	3.6	0.4	150	1150	7,580,000	226	0.0340	7,940,000	225	0.00829
7	3.6	0.5	300	1000	14,600,000	268	0.2310	13,200,000	267	0.01520
8	3.6	0.7	250	1050	12,400,000	252	0.0491	13,000,000	255	0.02290
9	4.0	0.3	250	1150	7,150,000	225	0.0363	7,500,000	223	0.01180
10	4.0	0.4	300	1100	8,770,000	217	0.0369	9,710,000	210	0.01170
11	4.0	0.5	150	1050	11,500,000	258	0.0447	12,500,000	257	0.01550
12	4.0	0.7	200	1000	14,600,000	282	0.0561	15,000,000	289	0.01760
13	4.4	0.3	300	1050	10,400,000	239	0.0425	10,200,000	220	0.00800
14	4.4	0.4	250	1000	12,200,000	293	0.0751	14,300,000	280	0.01140
15	4.4	0.5	200	1150	7,670,000	215	0.0278	8,140,000	209	0.00963
16	4.4	0.7	150	1100	10,300,000	252	0.0381	9,210,000	244	0.01190

Table 4 Design of experiments using L16 Orthogonal array for flashless forging

Exp. no	Billet temp. (°C)	Die temp. (°C)	Friction coeff.	Blocker			Finisher		
				Load (N)	Eff. stress (MPa)	Wear depth (mm)	Load (N)	Eff. stress (MPa)	Wear depth (mm)
1	1000	150	0.3	3,000,000	343	0.01370	2,850,000	210	0.00576
2	1000	200	0.4	3,200,000	253	0.01240	2,640,000	281	0.00439
3	1000	250	0.5	3,460,000	228	0.01430	2,790,000	234	0.00519
4	1000	300	0.7	3,870,000	253	0.00965	2,930,000	223	0.00478
5	1050	150	0.4	2,930,000	252	0.01200	2,410,000	212	0.00499
6	1050	200	0.3	2,660,000	241	0.01270	2,380,000	217	0.00487
7	1050	250	0.7	3,410,000	245	0.00879	2,650,000	227	0.00349
8	1050	300	0.5	3,000,000	227	0.01180	2,500,000	184	0.00383
9	1100	150	0.5	2,650,000	223	0.00953	2,330,000	196	0.00446
10	1100	200	0.7	2,990,000	219	0.00625	2,500,000	201	0.00359
11	1100	250	0.3	2,280,000	252	0.01030	2,070,000	198	0.00393
12	1100	300	0.4	2,430,000	187	0.01070	2,110,000	166	0.00409
13	1150	150	0.7	2,720,000	205	0.00517	2,040,000	187	0.00299
14	1150	200	0.5	2,280,000	187	0.00784	1,930,000	180	0.00413
15	1150	250	0.4	2,100,000	191	0.00751	1,790,000	173	0.00321
16	1150	300	0.3	1,950,000	201	0.00925	1,760,000	166	0.00344

applied to optimize all the parameters at once. And also are the output parameters being not of the same kind. So, a Grey–Taguchi method is a suitable technique, and the same is used in the present work for optimizing the output parameters. In the Grey–Taguchi method, all the parameters are initially converted into comparable grades. For analyzing the data in grey relation "lower the better" criteria is applied to all the output parameters. The output parameters are normalized using the following equation [9, 15–17].

$$x_i^*(k) = \frac{\max x_i(k) - x_i(k)}{\max x_i(k) - \min x_i^o(k)} \tag{3}$$

After getting the normalized outputs they are converted to grey relation coefficients. This y relation coefficient defines the relevancy between two different factors. For kth output in the *i*th experiment, the grey relation coefficient ξ_i(k) is given as follows [16–18].

$$\xi_i(k) = \frac{\Delta_{min} + \zeta\, \Delta_{max}}{\Delta_{0i} + \zeta\, \Delta_{max}} \tag{4}$$

where Δ_{0i}, is the deviation sequence of the comparability output with respect to reference output and, and is given by—$\Delta_{0i} = \|x_0(\mathrm{k}) - x_i(\mathrm{k})\|$, Δ_{max} and Δ_{min} are the minimum and maximum values of the deviation sequence. ζ is taken as 0.5.

Now the grey relational grades (GRG) is calculated as follows:

$$\gamma_i = \frac{1}{n}\sum_{k=1}^{n} w_k \xi_i(k) \tag{5}$$

where w_k is weightage of each factor, as equal weightage is given to all factors, w_k is taken as 1. The GRG is calculated for all the experiments and ranked in decreasing order of the grade. The first rank gives the best solution.

Grey relation grades are converted to S/N ratio as follows using larger the better criteria.

$$S/N = -10 \log\left(\sum (1/y^2)/n\right) \tag{6}$$

Then finally analysis of variance (ANOVA) is performed using the S/N ratio, to find out the most significant parameter affecting the output parameters at a 95% confidence level.

4 Results and Discussion

In forging operation, the load requirement to forge a component decides the capacity of the equipment to be used. And also the dies life is greatly affected by the amount of load coming on to the dies, if the load exceeds the permissible limit then the die will undergo a mechanical failure. Hence, the load should be as minimum as possible. The stress induced in the component during forging affects the functionality of the component. If the stress induced exceeds the permissible stress at a higher temperature, then the component cracks and has to be rejected. Hence, the stresses induced in the component should be as minimum as possible and also should be uniform all over the component. One of the most important factors that decides the durability of the dies is the amount of material removed due to wear. So the wear depth should be less. Owing to these reasons, "minimum is better" criteria used considered for optimization. To optimize the input parameters a Grey–Taguchi method is used as explained in Sect. 3. Normalized values, grey relation coefficient, grey relational grades, and ranks are calculated using the Eqs. (3–5) and tabulated in Tables 5 and 6 for conventional forging and flashless forging, respectively.

From Table 5, it is observed that the 15th run gave the best results. The input parameters for this run are Blocker flash thickness—4.4, friction coefficient—0.5, Die temperature—200 °C, Billet temperature—1150 °C. From Table 6 it is observed that the 15th run gave the best results. The input parameters for this run are Billet temperature—1150 °C, friction coefficient—0.4, Die temperature—250 °C. Grey relation grades are converted to S/N ratios using Eq. (6) and shown in Table 7. Which are further used for the analysis of variance. A validation simulation is carried out using the optimal parameters as inputs and the corresponding load distribution curve, effective stress, and wear depth are shown in Figs. 3 and 4 for conventional and flashless forging, respectively.

Next, analysis of variance for conventional and flashless forging is carried out at a confidence level of 95%, shown in Table 8 and Table 9 respectively. In case of conventional forging, it is found that the p-value of billet temperature is less than 0.05, predicting that billet temperature is the most significant parameter which affects the load, effective stress, and wear depth. In the case of flashless forging, predicting that billet temperature and die temperature are the most significant parameters which affect the load, effective stress, and wear depth.

5 Conclusion

- The present study presents a complete die design and optimization procedure for the flashless forging of a coupling flange.
- It also presents a comparative study between flashless forging and conventional forging process.

Table 5 Calculated normalized values, grey relational coefficient, and grey relational grade (GRG) for conventional forging

Exp. no	Normalized outputs						Grey relation coefficient						GRG	Rank
	Blocker			Finisher			Blocker			Finisher				
	Load	Eff. stress	Wear depth	Load	Eff. stress	Wear depth	Load	Eff. stress	Wear depth	Load	Eff. stress	Wear depth		
1	0.000	0.000	0.000	0.413	0.020	0.705	0.333	0.333	0.333	0.460	0.338	0.629	0.404	16
2	0.579	0.544	0.920	0.453	0.496	0.889	0.543	0.523	0.862	0.478	0.498	0.818	0.620	10
3	0.801	0.746	0.935	0.640	0.836	0.875	0.715	0.663	0.884	0.581	0.753	0.800	0.733	7
4	0.770	0.922	0.953	0.827	0.992	0.913	0.685	0.865	0.914	0.743	0.985	0.851	0.841	4
5	0.786	0.641	0.956	0.748	0.681	0.852	0.700	0.582	0.920	0.665	0.610	0.772	0.708	8
6	0.953	0.862	0.980	0.941	0.804	0.981	0.914	0.784	0.962	0.895	0.718	0.963	0.873	2
7	0.186	0.372	0.349	0.240	0.273	0.517	0.380	0.443	0.434	0.397	0.408	0.509	0.429	15
8	0.426	0.561	0.932	0.267	0.422	0.000	0.466	0.533	0.880	0.405	0.464	0.333	0.513	12
9	1.000	0.880	0.973	1.000	0.827	0.745	1.000	0.807	0.948	1.000	0.743	0.662	0.860	3
10	0.823	0.971	0.971	0.705	0.998	0.752	0.738	0.945	0.945	0.629	0.995	0.668	0.820	5
11	0.525	0.488	0.946	0.333	0.400	0.497	0.513	0.494	0.902	0.429	0.455	0.498	0.548	11
12	0.186	0.209	0.909	0.000	0.000	0.356	0.380	0.387	0.847	0.333	0.333	0.437	0.453	14
13	0.645	0.718	0.953	0.640	0.863	1.000	0.585	0.639	0.914	0.581	0.785	1.000	0.751	6
14	0.448	0.085	0.848	0.093	0.114	0.772	0.475	0.353	0.767	0.355	0.361	0.687	0.499	13
15	0.943	1.000	1.000	0.915	1.000	0.891	0.898	1.000	1.000	0.854	1.000	0.820	0.929	1
16	0.656	0.560	0.967	0.772	0.564	0.738	0.592	0.532	0.938	0.687	0.534	0.656	0.657	9

Table 6 Calculated normalized values, grey relational coefficient, and grey relational grade (GRG) for flashless forging

Exp. no	Normalized outputs						Grey relation coefficient						GRG	Rank
	Blocker			Finisher			Blocker			Finisher				
	Load	Eff. stress	Wear depth	Load	Eff. stress	Wear depth	Load	Eff. stress	Wear depth	Load	Eff. stress	Wear depth		
1	0.453	0.000	0.066	0.068	0.617	0.000	0.478	0.333	0.349	0.349	0.566	0.333	0.401	16
2	0.349	0.576	0.208	0.248	0.000	0.495	0.434	0.541	0.387	0.399	0.333	0.497	0.432	14
3	0.214	0.736	0.000	0.120	0.403	0.206	0.389	0.654	0.333	0.362	0.456	0.386	0.430	15
4	0.000	0.575	0.509	0.000	0.500	0.354	0.333	0.541	0.505	0.333	0.500	0.436	0.441	13
5	0.490	0.584	0.252	0.444	0.601	0.278	0.495	0.546	0.401	0.474	0.556	0.409	0.480	12
6	0.630	0.654	0.175	0.470	0.556	0.321	0.575	0.591	0.377	0.485	0.530	0.424	0.497	11
7	0.240	0.627	0.604	0.239	0.468	0.819	0.397	0.573	0.558	0.397	0.485	0.735	0.524	10
8	0.453	0.745	0.274	0.368	0.837	0.697	0.478	0.662	0.408	0.442	0.754	0.622	0.561	9
9	0.635	0.770	0.522	0.513	0.732	0.471	0.578	0.685	0.511	0.506	0.651	0.486	0.570	8
10	0.458	0.795	0.882	0.368	0.694	0.783	0.480	0.709	0.809	0.442	0.620	0.698	0.626	6
11	0.828	0.582	0.438	0.735	0.722	0.661	0.744	0.544	0.471	0.654	0.643	0.596	0.609	7
12	0.750	0.999	0.394	0.701	1.000	0.603	0.667	0.998	0.452	0.626	1.000	0.557	0.717	5
13	0.599	0.886	1.000	0.761	0.817	1.000	0.555	0.814	1.000	0.676	0.732	1.000	0.796	3
14	0.828	1.000	0.708	0.855	0.873	0.588	0.744	1.000	0.631	0.775	0.798	0.549	0.749	4
15	0.922	0.974	0.744	0.974	0.934	0.921	0.865	0.950	0.661	0.951	0.884	0.863	0.862	1
16	1.000	0.912	0.553	1.000	0.995	0.838	1.000	0.851	0.528	1.000	0.991	0.755	0.854	2

Table 7 S/N rations for grey relation grades (GRG)

Conventional forging			Flashless forging		
Exp. no	GRG	S/N ratio	Exp. no	GRG	S/N ratio
1	0.404	−7.87237	1	0.401	−7.92843
2	0.623	−4.11024	2	0.432	−7.28871
3	0.735	−2.67425	3	0.430	−7.32888
4	0.843	−1.48345	4	0.441	−7.10307
5	0.711	−2.96261	5	0.480	−6.37350
6	0.876	−1.14992	6	0.497	−6.06958
7	0.429	−7.35085	7	0.524	−5.61563
8	0.516	−5.74701	8	0.561	−5.02257
9	0.863	−1.27978	9	0.570	−4.88543
10	0.823	−1.69200	10	0.626	−4.06545
11	0.551	−5.17697	11	0.609	−4.31392
12	0.455	−6.83977	12	0.717	−2.89332
13	0.753	−2.46410	13	0.796	−1.97877
14	0.502	−5.98593	14	0.749	−2.50613
15	0.929	−0.63969	15	0.862	−1.28538
16	0.659	−3.62229	16	0.854	−1.36953

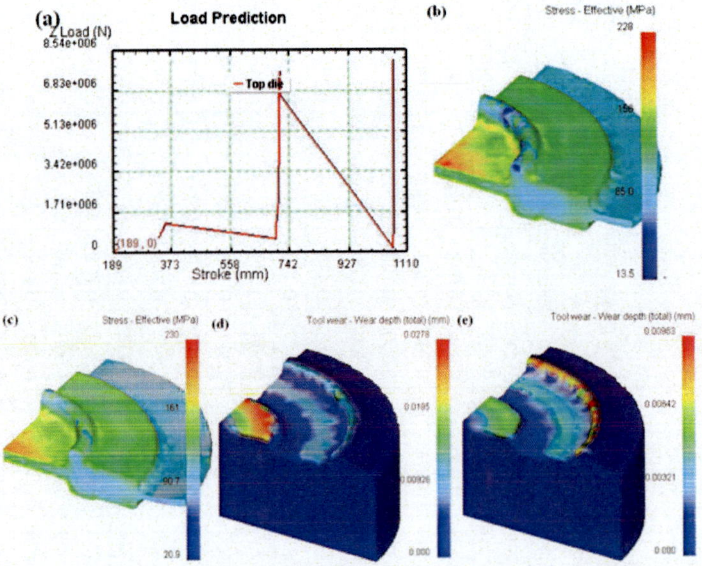

Fig. 3 Simulation results for conventional forging **a** Load prediction curve **b** Effective stress for blocker **c** Blocker wear depth **d** Effective stress for finisher **e** Finisher wear depth

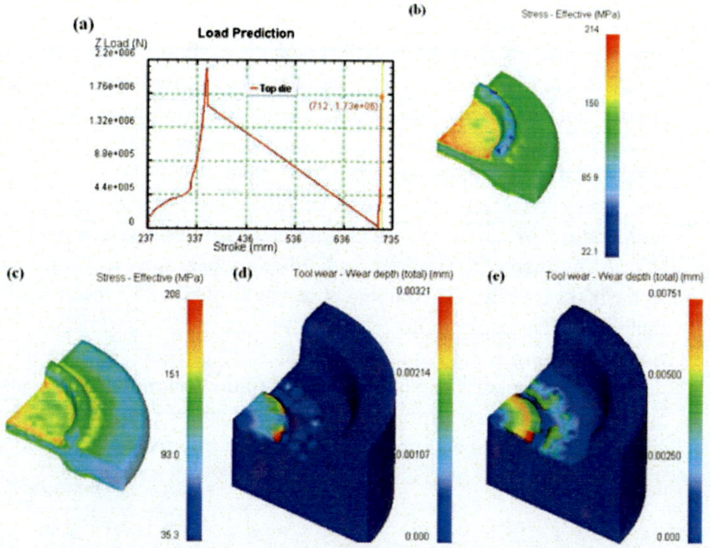

Fig. 4 Simulation results for flashless forging **a** Load prediction curve **b** Effective stress for blocker **c** Effective stress for finisher **d** Blocker wear depth **e** Finisher wear depth

Table 8 Analysis of variance using grey relation grades for conventional forging

Source	DF	Adj SS	Adj MS	F-value	P-value	Remarks
Flash thickness	3	2.786	0.9287	1.75	0.328	Insignificant
Friction coefficient	3	3.028	1.0092	1.90	0.305	Insignificant
Die temperature	3	3.099	1.0330	1.95	0.299	Insignificant
Billet temperature	3	75.444	25.1480	47.44	0.005	Significant
Error	3	1.590	0.5301			
Total	15	85.947				

Table 9 Analysis of variance using grey relation grades for flashless forging

Source	DF	Adj SS	Adj MS	F-value	P-value	Remarks
Billet temp. (c)	3	69.7001	23.2334	184.48	0.000	Significant
Die temp.(c)	3	3.1462	1.0487	8.33	0.015	Significant
Friction coeff.	3	0.6040	0.2013	1.60	0.286	Insignificant
Error	6	0.7557	0.1259			
Total	15	74.2059				

- Grey–Taguchi method a multi-objective optimization technique is used to optimize the process parameters—Flash thickness, billet temperature, friction coefficient & die temperature in case of conventional forging and billet temperature, friction coefficient and die temperature in case of flashless forging.
- Analysis of variance (ANOVA) at 95% confidence level is carried out to find the most significant input parameter in both cases.
- The most significant factor that affects the conventional forging process is billet temperature. Whereas, in case of flashless forging die temperature also came out to be a significant factor along with billet temperature.
- In flashless forging, the component can be forged with a lesser load compared to that of conventional forging.
- Since trimming is not required in flashless forging, a complete set for trimming die, trimming punch and equipment is completely eliminated. And also, grain flow is not outcropped as a result of which mechanical properties of the component like fatigue strength and impact strength also improved.

References

1. Jong-Heon L-H, Bae K-B (1997) A study on flash- and flashless-precision forging by the upper-bound elemental technique. J Mater Process Technol 72:371–379
2. Vazquez V, Altan T (2000) Die design for flashless forging of complex parts. J Mater Process Technol 98:81–89
3. Li F, Chena P, Hana J, Denga L, Yi J, Liu Y, Eckert J (2020) Metal flow behavior of P/M connecting rod preform in flashless forging based on isothermal compression and numerical simulation. J Market Res 9(2):1200–1209
4. Duggirala R (1989) Design of forging dies for forming flashless ring gear blanks using finite element methods. J Mater Shaping Technol 7:33–47
5. Park H-S, Risky F, Kumar S (2018) Preform optimization for bevel gear of warm forging process. Procedia CIRP 72:340–345
6. Takemasu T, Vazquez V, Painter B, Altan T (1996) Investigation of metal flow and preform optimization in flashless forging of a connecting rod. J Mater Process Technol 59:95–105
7. Yanhui Y, Dong L, Ziyan H, Zijian L (2009) Multi-objective preform optimization using RSM. Rare Met Mater Eng 38
8. Huang Y, Du CK, Wang LG, Zhang ZY, Zhuang XW, Liu Q (2019) Multi-objective preform optimization for spherical hinge mandrel based on response surface methodology. In: IOP conference series: materials science and engineering, vol 504. IEEE, p 012050
9. Tosun N, Pihtili H (2010) Grey relational analysis of performance characteristics in MQL milling of 7075 Al alloy. Int J Adv Manuf Technol 46:509–515
10. Feng W, Hua L (2011) Multi-objective optimization of process parameters for the helical gear precision forging by using Taguchi method. J Mech Sci Technol 25(6):1519–1527
11. Pal S, Malviya SK, Pal SK, Samantaray AK (2009) Optimization of quality characteristics parameters in a pulsed metal inert gas welding process using grey-based Taguchi method. Int J Adv Manuf Technol 44:1250–1260
12. Zhang J, Wu D, Zhou J, Wang J (2014) Multi-objective optimization of process parameters for 7050 aluminum alloy rib-web forgings' precise forming based on Taguchi method, Procedia Eng 81:558–563
13. Raj MP, Kumar M, Pramanick AK (2020) Yield improvement in hot forging of differential spider. Mater Today Proc 26:3107–3115. (Part 2)

14. Thomas A (1995) Die design, forging handbook. Drop Forging Research Association, Sheffield, UK
15. Haq AN, Marimuthu P, Jeyapaul R (2008) Multi response optimization of machining parameters of drilling Al/SiC metal matrix composite using grey relational analysis in the Taguchi method. Int J Adv Manuf Technol 37(3–4):250–255
16. Tzeng CJ, Lin YH, Yang YK, Jeng MC (2009) Optimization of turning operations with multiple performance characteristics using the Taguchi method and Grey relational analysis. J Mater Process Technol 209(6):2753–2759
17. Lin ZC, Ho CY (2003) Analysis and application of grey relation and ANOVA in chemical–mechanical polishing process parameters. Int J Adv Manuf Technol 21:10–14. https://doi.org/10.1007/s001700300001
18. IS 3469-1 to 3 (1974) Tolerances for closed die steel forgings [MTD 16: Alloy Steels and Forgings]

Investigation of Frequency Analysis of Functionally Graded Plate Under Thermal Effect with the Help of FEM

Pankaj Sharma, Ashish Khinchi, and Monika Meena

1 Introduction

The recent development in design and manufacturing technologies have greatly enhanced the use of functionally graded plate structures in a wide range of engineering fields, i.e., space vehicles, nuclear reactors, automobile industry, and other high-temperature application fields [1, 7].

FG materials (FGMs) are advanced materials and are being examined for use in structures subjected to the thermal environments. FGM constituents are designed in such a way that composite material limitations are overcome. Functionally graded materials have various kinds of benefits such as temperature resistance properties, wear resistance, reduction in stress concentration, elimination of thermal stresses, and an increase in strength to weight ratio. This is possible due to the material phases of a functionally graded material changing gradually through a certain direction. Due to the excellent material properties of FGMs, they are used in many engineering fields, i.e., aerospace, defence, space shuttle, turbine blades, rocket engine components, electronics and biomedical fields [2–4].

Many types of research have been published about FGM plates for specific work. Shahidzadeh et al. [5] dealt with the FE approach for the harmonic study of functionally graded plates. Power law is employed in this work. Material parameters are graded in the thickness direction. Ramu and Mohanty [6] dealt with harmonic behavior of isotropic rectangular plates using the FEM approach. Ramu and Mohanty [7] also presented the modal behavior of FG-rectangular plate using the FEM

P. Sharma · M. Meena
Department of Mechanical Engineering, Rajasthan Technical University, Kota, Rajasthan, India
e-mail: psharma@rtu.ac.in

A. Khinchi (✉)
Department of Mechanical Engineering, Vedant College of Engineering and Technology, Kota, Rajasthan, India

R. Prasad et al. (eds.), *Advancement in Materials Processing Technology*,
Springer Proceedings in Materials 12,
https://doi.org/10.1007/978-981-16-3297-6_5

approach. In this analysis, material constants are varied according to power law in the z-axis direction. Wenjie et al. [8] dealt with modal analysis of different shape isotropic plates. In this analysis, different kinds of boundary conditions are used by the author. The harmonic characteristics of functionally graded plates with the help of HSDT theory were investigated by Talha et al. [9]. Material constants are varying in the direction of thickness using P-FGM law. Ramu et al. [10] investigated the harmonic behavior of functionally graded plates with hygrothermal effects. Material parameters are temperature and moisture dependent. The material constants are varying in the direction of thickness. The modal study of FGM rectangular plate subjected to the thermal environment was presented by lie et al. [11]. Material parameters are a function of temperature and are varying in the direction of thickness. Modal study of FG-plate with different cut-out under thermal effect was presented by Maziar et al. [12]. Material constants are varied according to power law in the thickness direction. Effect of several factors, i.e., cut out effect, loading effect, and effect of boundary conditions on eigenfrequencies are presented. Swaminathan et al. [13] presented a detailed review of FGM plate under thermal effect. The detailed analysis on functionally graded structure with and without thermal effect in a certain direction using COMSOL are available in the literature [14–17].

In most of the research, material constants are varying in thickness direction. In this research, material constants of FGM rectangular plate are graded in the longitudinal direction. Material parameters are temperature dependent. Material constants are varying in the z-axis according to power law gradation. From the literature, it can be observed that modal characteristics of a functionally graded rectangular plate in the axial direction under the thermal effect were never analyzed.

The aim of this research is to obtain vibration characteristics for square FGM plates in the axial direction. Mode shapes of the FGM plate have been obtained by using the FEM software (COMSOL). The material constants of the FGM plate are a function of temperature and are supposed to be continuously varied in longitudinal axis according to power law gradation. A modal study of an FGM square plate in the axial direction has been performed by COMSOL (version 5.5).

2 Theory and Formulation

A square plate has geometrical parameters, i.e., side (a), thickness (H). A square plate is made of functionally graded material. FGM consist of the ceramic part (Si_3N_4) and metal part (SUS_3O_4) as shown in Fig. 1. Material parameters of FGM plate are taken from the Sharma et al. [14].

In this study, effective material constants of the functionally graded plate are varied continuously in the longitudinal axis (X-axis) according to power law gradation and a function of temperature (T) equation [13] as follows:

$$E_{z,T} = (E_c - E_m)V_f^n + E_m \tag{1}$$

Fig. 1 FGM plate

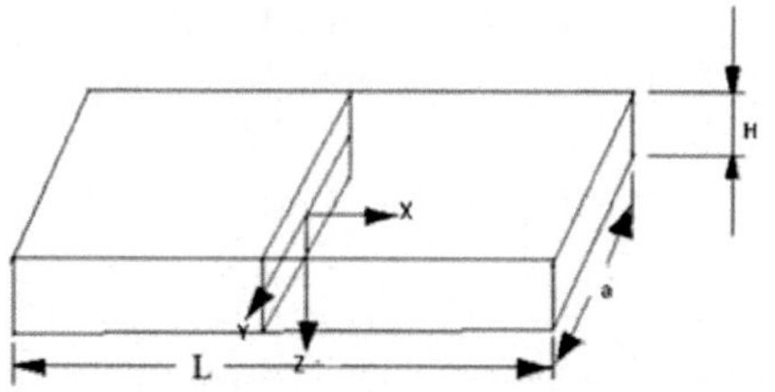

$$v_{z,T} = (v_c - v_m)V_f^n + v_m \quad (2)$$

$$\rho_{z,T} = (\rho_c - \rho_m)V_f^n + \rho_m \quad (3)$$

$$K_{z,T} = (K_c - K_m)V_f^n + K_m \quad (4)$$

$$\alpha_{z,T} = (\alpha_c - \alpha_m)V_f^n + \alpha_m \quad (5)$$

$$V_f = \left(\frac{x}{h} + \frac{1}{2}\right) \quad (6)$$

$$L(T) = L_0\left(L_{-1}T^{-1} + L_1T^1 + L_2T^2 + L_3T^3\right) \quad (7)$$

where the modulus of elasticity, Poisson's ratio, density, thermal conductivity, volume fraction of the ceramic, and volume fraction index are E, v, ρ, K, $V_{f,}$ and n, respectively. L_0, L_{-1}, L_1, L_2, and L_3 are coefficients. These are functions of temperature for ceramic and metal materials. Subscripts are c (ceramic) and m (metal).

Displacement field

According to classical plate theory, displacement fields [7] are as follows:

$$U = u(x, y) - z \cdot w_x(x, y) \quad (8)$$

$$V = u(x, y) - z \cdot w_y(x, y) \quad (9)$$

$$W = w(x, y) \quad (10)$$

Here U and V are displacement in the x and y direction, respectively. W is the displacement, function of x and y. Notations are used in the equations are $\frac{\partial f}{\partial x} = f_x$, $\frac{\partial^2 f}{\partial x^2} = f_{xx}$ and $\frac{\partial^2 f}{dy\partial x} = f_{xy}$.

Strain–displacement relation

$$e_x = u_x - zw_{xx} \tag{11}$$

$$e_y = v_y - zw_{yy} \tag{12}$$

$$\Upsilon_{xy} = u_y + v_x - 2zw_{xy} \tag{13}$$

$$\Upsilon_{yz} = 0; \Upsilon_{xz} = 0 \tag{14}$$

Here, $e_{x,}$ e_y and Υ are normal and shear strains, respectively.

Constitutive equations

The state of stress in the global x–y−z coordinate system of the functionally graded plate can be written as

$$\begin{Bmatrix} \sigma_{xx} \\ \sigma_{yy} \\ \tau_{xy} \end{Bmatrix} = \begin{bmatrix} Q_{11} & Q_{12} & 0 \\ Q_{21} & Q_{22} & 0 \\ 0 & 0 & Q_{66} \end{bmatrix} \begin{Bmatrix} e_x \\ e_y \\ \Upsilon_{xy} \end{Bmatrix} \tag{15}$$

σ_{xx}, $\sigma_{yy,}$ and τ_{xy} are normal and shear stress, respectively.

$$Q_{11} = E(x)/\left(1 - v^2\right) \tag{16}$$

$$Q_{12} = Q_{21} = vE(x)/\left(1 - v^2\right) \tag{17}$$

$$Q_{66} = E(x)/2(1 + v) \tag{18}$$

Variational form [14]

The total strain energy of the plate can be given as

$$S_e = \int\limits_V \delta\left(\sigma_{xx} e_x + \sigma_{yy} e_y + \tau_{xy} \Upsilon_{xy}\right) \mathrm{dv} \tag{19}$$

The kinetic energy can be expressed as

$$K_e = \frac{1}{2} \iint\limits_A \rho h w^2 \mathrm{dA} \tag{20}$$

Work done (W) due to thermal load expressed as

$$W_e = -\frac{1}{2}\int_0^a N^t w^2 dx \tag{21}$$

$$N^t = \int_{\frac{-h}{2}}^{\frac{h}{2}} Q_{11}(\mathrm{z})\,\alpha(\mathrm{z})\big[T(z) - T^0\big]adz \tag{22}$$

Equation of motion for plate element

$$\delta\int_{t_1}^{t_2} (S_e - K_e + W_e)dt = 0 \tag{23}$$

Governing equation of motion

$$\big([K] - \omega^2[M]\big)\{w\} = 0 \tag{24}$$

Thermal effect

Here, a uniform temperature load along the axial direction of the FGM plate is considered. The uniform temperatures equation is as follows [13].

$$\mathrm{T} = \Delta\mathrm{T} + \mathrm{T}_0 \tag{25}$$

3 Comprasion Study

Tables (1, 2, 3, 4 and 5).

4 Parametric Study

In the parametric study, the eigenfrequencies and the vibrational mode of the axially square FG plate are calculated and presented. The axially square FG plate considered in the parametric study is made of SUS_3O_4 and Si_3N_4, respectively. In the other case, this plate is also subjected to a thermal environment. In this study, the reference temperature of the axially functionally graded square plate is supposed to be $T_0 = 300$ K. It is considered that the side of the plate is a and thickness is H. The effects of aspect ratios (Φ), volume fraction indices (n), and temperature rise (ΔT) on natural frequencies are studied under thermal effect (Tables 6, 7, 8 and 9, Figs. 2, 3, 4, 5 and 6).

Table 1 Natural frequencies for isotropic rectangular plate (a = 0.6 m, b = 0.4 m, and h = 0.00625 m) under simply supported boundary condition (SSSS) using COMSOL [6]

Mode	Ref.	COMSOL
1	136.5	136.45
2	262.6	262.19
3	420.1	419.09
4	472.7	471.32
5	546.2	544.39
6	756.35	752.77
7	766.85	763.18
8	892.9	887.93
9	1018.9	1012.5
10	1050.4	1043.6

Table 2 Non-dimensional frequencies parameter $\left(\overline{\omega} = \omega a^2\sqrt{\frac{\rho h}{D}}\right)$ for isotropic plate (a = 1 m, b = 1 m and h = 0.02 m) under clamped boundary (CCCC) condition using COMSOL [8]

Mode	Ref.	COMSOL
1	35.952	35.986
2	73.318	73.232
3	73.326	73.258
4	108.10	107.772
5	131.44	130.95
6	132.00	131.590

Table 3 Non-dimensional frequencies parameter $\left(\overline{\omega} = {\omega \alpha^2}/{h}\sqrt{{\rho_c}/{E_c}}\right)$ for simply supported (SSSS) FGM (SUS_3O_4/Si_3N_4) plate (a = b = 0.1 m, h = 0.01 m) under power law using COMSOL [9]

Mode	n = 0.5		n = 1		n = 5		n = 10	
	Ref.	COMSOL	Ref.	COMSOL	Ref.	COMSOL	Ref.	COMSOL
1	3.9701	4.1228	3.4845	3.5716	2.8351	2.5548	2.6973	2.5082
2	9.6890	9.8386	8.4903	8.48605	6.8941	6.0825	6.5669	5.9710
3	9.6906	9.8386	8.4918	8.48605	6.8952	6.0825	6.5680	5.9710
4	14.9404	15.1010	13.0959	13.366	10.6102	9.3186	10.1053	9.1267
5	18.7691	18.4018	16.4526	15.7822	13.3057	11.3427	12.6668	11.1331

Table 4 Non-dimensional frequencies parameter $\left(\overline{\omega} = {}^{\omega\alpha^2}/_{h}\sqrt{{}^{\rho_c}/_{E_c}}\right)$ for clamped (CCCC) FGM (SUS_3O_4/Si_3N_4) square plate (a = b = 0.1 m h = 0.01 m) under power law using COMSOL [9]

Mode	n = 0.5		n = 1		n = 5		n = 10	
	Ref.	COMSOL	Ref.	COMSOL	Ref.	COMSOL	Ref.	COMSOL
1	7.0202	6.8373	6.1489	5.9981	4.9816	4.8817	4.7457	4.6598
2	13.7978	13.1138	12.0812	11.4904	9.7440	9.3061	9.2841	8.8826
3	13.7978	13.1154	12.0812	11.4915	9.7440	9.3072	9.2841	8.8837
4	19.4845	18.4436	17.0625	16.1489	13.7350	13.0421	13.0873	12.4473
5	23.9945	21.7917	20.9992	19.0719	16.8507	15.3703	16.0556	14.6699

Table 5 Comparisons of first-dimensionless frequencies $\left(\overline{\omega} = {}^{\omega\alpha^2}/_{h}\sqrt{{}^{I_0}/_{D_0}}\right)$ for Clamped FGM (Si_3N_4/SUS_30_4) plates with thermal effect (L = 0.2 m, H/W = 0.1 and T_{ref} = 300 K)

ΔT (K)	Source	Ref.	COMSOL
500	Li et al. [11]	3.2357	3.1116
300	Ramu et al. [10]	3.6618	3.1829
0	Li et al. [11]	4.1658	3.2467

Table 6 The natural frequencies of axially FG square plate for CCCC and SSSS end conditions (a = b = 0.1 m, H = 0.01 m) under without thermal effect

Mode	Natural frequencies (Hz)	
	CCCC	SSSS
1	17,999	10,645
2	32,117	23,965
3	35,678	27,396
4	49,558	41,307

Table 7 The natural frequencies of axially FG square plate versus uniform temperature rise for clamped boundary conditions (CCCC) under thermal effect (a = b = 0.2 m, H = 0.02 m, and T_{ref} = 300 K)

ΔT	Mode 1	Mode 2	Mode 3	Mode 4	Mode 5
500	4811.2	9217.3	9220.3	12,956	15,312
300	4840.8	9274	9277.1	13,036	15,407
0	4895.2	9378.1	9381.2	13,182	15,580

Table 8 The natural frequencies of axially FG square plate v/s volume fraction index for clamped boundary conditions (CCCC) under uniform temperature (a = b = 0.2, H = 0.02, and ΔT (K) = 300 K)

Mode	n = 1	n = 2	n = 3	n = 4	n = 5	n = 6
1	5729.9	4840.8	4455.6	4265.1	4165.3	4111.5
2	10,991	9274.1	8529.6	8161.5	7968.7	7864.7
3	10,995	9277.1	8532.4	8164.2	7971.3	7867.3
4	15,463	13,036	11,983.00	11,463	11,190	11,043
5	18,284	15,407	14,159	13,542	13,219	13,045

Table 9 The natural frequencies of axially FG square plate v/s aspect ratio (Φ) for clamped boundary conditions (CCCC) under thermal effect

Mode	Φ = 10	Φ = 15	Φ = 20	Φ = 25	Φ = 30
1	4840.5	2280.7	1311.1	842.33	586.06
2	9274	4517.3	2637	1702.5	1188.00
3	9277.1	4533.7	2643.7	1704.5	1189.1
4	13,036	6515.2	3844.6	2490.1	1742.6

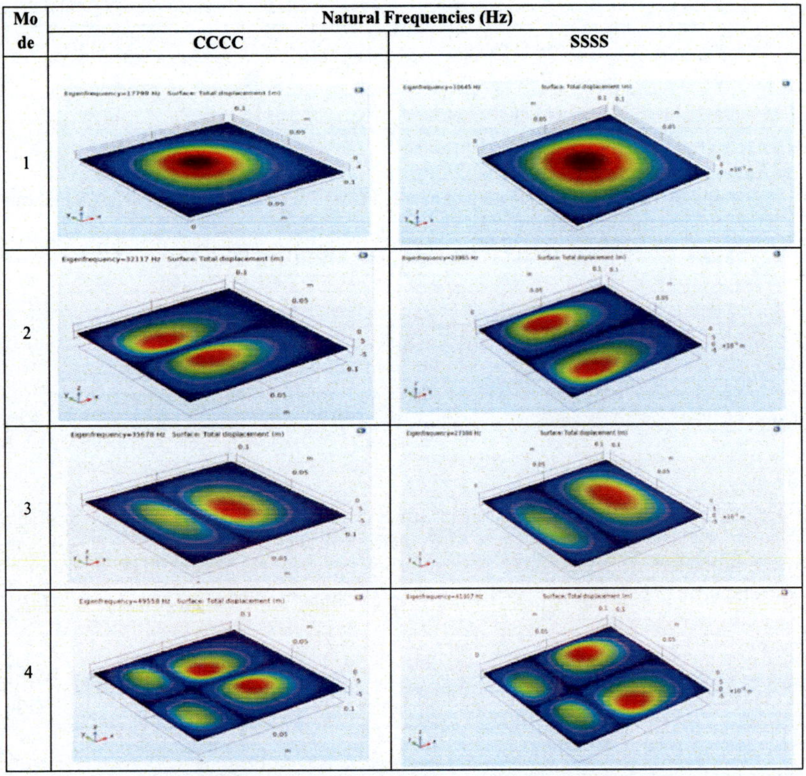

Fig. 2 Vibration mode of axially FG square plate for CCCC and SSSS end conditions (a = b = 0.1 m, H = 0.01 m) under without thermal effect

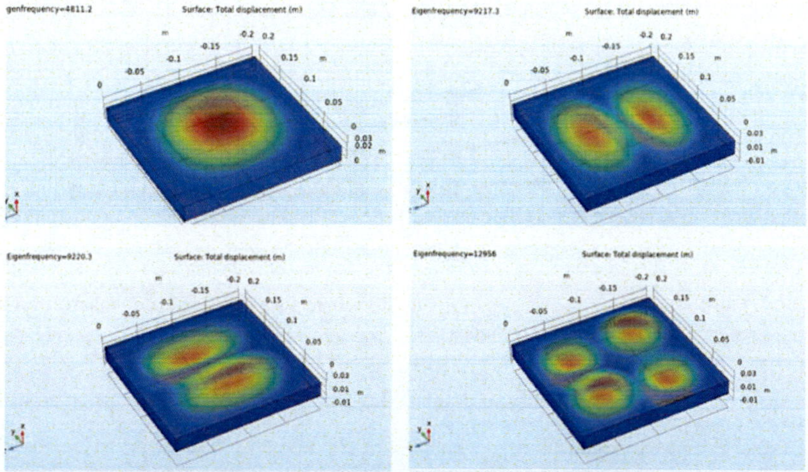

Fig. 3 First four vibration mode of axially FG square plate for clamped boundary conditions (CCCC) under thermal effect (a = b = 0.2 m, H = 0.02 m, and ΔT = 500 K)

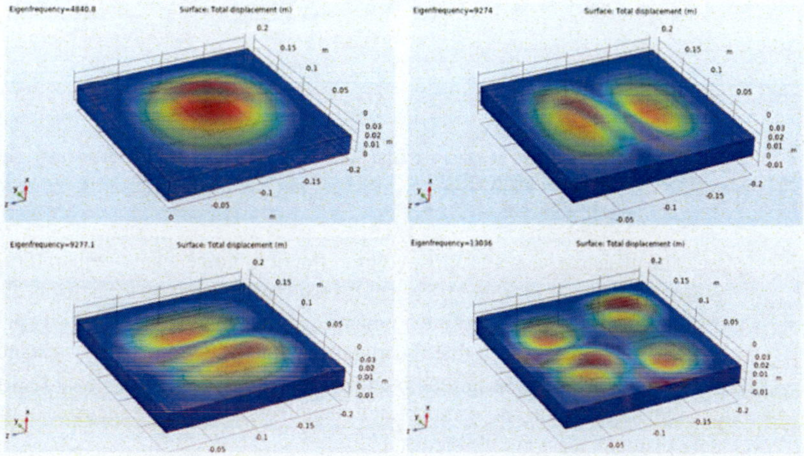

Fig. 4 First four vibration mode of axially FG square plate for clamped boundary conditions (CCCC) under thermal effect (a = b = 0.2 m, H = 0.02 m, and ΔT = 300 K)

5 Conclusions

The modal study of the functionally graded square plate is used in this research. The material parameters have continuous variation in the longitudinal axis according to power law. COMSOL-5.5 is employed to obtain eigenfrequencies of the FGM

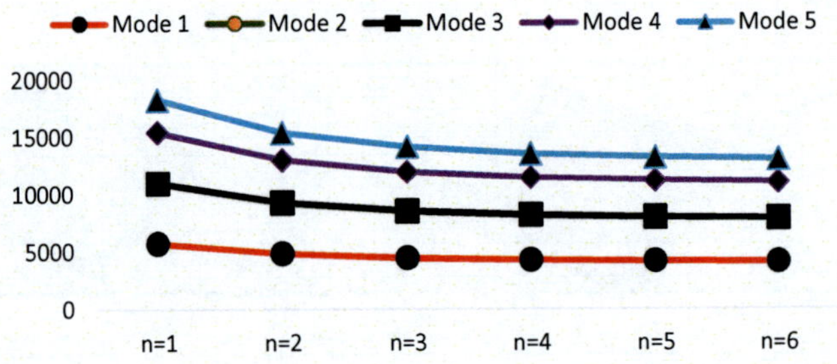

Fig. 5 The natural frequencies of axially FG square plate for variation of volume fraction index under clamped boundary conditions (CCCC)

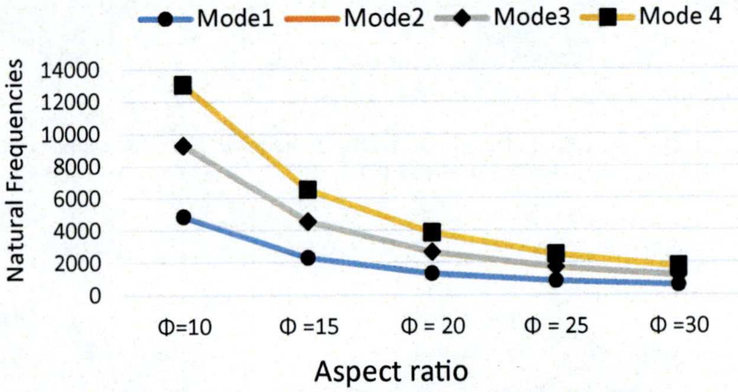

Fig. 6 Effect of aspect ratio on natural frequencies for FGM square plate with thermal effect under clamped boundary condition (CCCC)

plate. The effects of aspect ratios, volume fraction indices, and temperature rise on eigenfrequencies are presented. The study results obtained with the help of COMSOL (version 5.5) are in good agreement with the results obtained in previously published work. It has been analyzed that the eigenfrequencies of functionally graded plates decrease with an increase in temperature rise. It may also be revealed that with an increase in aspect ratio and volume fraction indices, the eigenfrequencies of the FGM square plate decrease for all modes.

References

1. Sharma P (2019) Vibration analysis of functionally graded piezoelectric actuators. Springer, New York, NY. https://doi.org/10.1007/978-981-13-3717-8
2. Khinchi A, Sharma P (2020) Free vibration analysis of isotropic spherical cap and FG-spherical cap with cut-out using COMSOL. In: AIP conference proceedings, vol 2220 (no. 1). AIP Publishing LLC, p 130074. https://doi.org/10.1063/5.0001299

3. Singh R, Sharma P (2019). A review on modal characteristics of FGM structures. In: AIP conference proceedings, vol 2148 (no. 1). AIP Publishing LLC, p 030037. https://doi.org/10.1063/1.5123959
4. Sharma P, Parashar SK (2016) Free vibration analysis of shear-induced flexural vibration of FGPM annular plate using generalized differential quadrature method. Compos Struct 155:213–222. https://doi.org/10.1016/j.compstruct.2016.07.077
5. Tabatabaei SS, Fattahi AM (2020) A finite element method for modal analysis of FGM plates. Mech Based Des Struct Mach 1–12. https://doi.org/10.1080/15397734.2020.1744004
6. Ramu I, Mohanty SC (2012) Study on free vibration analysis of rectangular plate structures using finite element method. Procedia Eng 38:2758–2766. https://doi.org/10.1016/j.proeng.2012.06.323
7. Ramu I, Mohanty SC (2014) Modal analysis of functionally graded material plates using finite element method. Procedia Mater Sci 6:460–467. https://doi.org/10.1016/j.mspro.2014.07.059
8. Guo W, Feng Q (2019) Free vibration analysis of arbitrary-shaped plates based on the improved rayleigh–ritz method. Adv Civ Eng. https://doi.org/10.1155/2019/7041592
9. Talha M, Singh BN (2010) Static response and free vibration analysis of FGM plates using higher order shear deformation theory. Appl Math Model 34(12):3991–4011. https://doi.org/10.1016/j.apm.2010.03.034
10. Ramu I, Narendra M, Venu M (2018) Effect of hygrothermal environment on free vibration characteristics of FGM plates by finite element approach. In: International conference on mechanical, materials and renewable energy. IOP conference series: materials science and engineering, vol 377 (no. 1). https://doi.org/10.1088/1757-899X/377/1/012021
11. Li Q, Iu VP, Kou KP (2009) Three-dimensional vibration analysis of functionally graded material plates in thermal environment. J Sound Vib 324(3–5):733–750. https://doi.org/10.1016/j.jsv.2009.02.036
12. Janghorban M, Zare A (2011) Thermal effect on free vibration analysis of functionally graded arbitrary straight-sided plates with different cutouts. Lat Am J Solids Struct 8(3):245–257. https://doi.org/10.1590/S1679-78252011000300003
13. Swaminathan K, Sangeetha DM (2017) Thermal analysis of FGM plates–a critical review of various modeling techniques and solution methods. Compos Struct 160:43–60. https://doi.org/10.1016/j.compstruct.2016.10.047
14. Sharma P, Singh R, Hussain M (2020) On modal analysis of axially functionally graded material beam under hygrothermal effect. Proc Inst Mech Eng C J Mech Eng Sci 234(5):1085–1101. https://doi.org/10.1177/0954406219888234
15. Sharma P, Singh R (2019) Investigation on modal behaviour of FGM annular plate under hygrothermal effect. In: IOP conference series: materials science and engineering, vol 624, (no. 1). IOP Publishing, p 012001. https://doi.org/10.1088/1757-899X/624/1/012001
16. Parashar SK, Sharma P (2016) Modal analysis of shear-induced flexural vibration of FGPM beam using generalized differential quadrature method. Compos Struct 139:222–232. https://doi.org/10.1016/j.compstruct.2015.12.012
17. Sharma P (2018) Efficacy of harmonic differential quadrature method to vibration analysis of FGPM beam. Compos Struct 189:107–116. https://doi.org/10.1016/j.comp-struct.2018.01.059

Numerical Prediction of Residual Stress and Temperature Field in TIG Welding

A. Anandhu, S. R. Sarath, B. P. Pabin, H. B. Mohammed Nazim, V. M. Varma Prasad, and R. Ranju

1 Introduction

Welded joints are widely used in the fabrication industry including shipbuilding, aerospace, pressure vessels, and steel bridges. There are several types of welding techniques. Tungsten inert gas welding, stick welding, gas metal arc welding, and flux core arc welding are the types of welding used commonly.

A major problem in welding is the development of residual stress. Residual stresses arise due to no external loading on welded portions. The effects produced by residual stresses are cracking of material, distortion tenancy, and decreased strength of the material. This bought with it the need to reduce the residual stress during welding. So that post-weld treatment, design considerations, vibratory stress relieving were developed to reduce it.

In most of the mechanical structures, piping systems are made using stainless steel and mild steel. These structures were fabricated or joined by using arc welding methods such as carbon arc welding and TIG welding.

TIG welding is commonly used for joining stainless steel and mild steel structures because while comparing other welding techniques TIG welded joints have more improved performance.

The objective of this paper is to analyze temperature field distribution and residual stress present in the stainless steel and mild steel welded joints using FEM with an accepted numerical model. The determination of numerical models will help to optimize the welding parameters for enhancing the weld strength.

A. Anandhu (✉) · S. R. Sarath · B. P. Pabin · H. B. M. Nazim · V. M. V. Prasad · R. Ranju
Mechanical Department, Musaliar College of Engineering, Chirayinkeezhu, Thiruvananthapuram, India

R. Prasad et al. (eds.), *Advancement in Materials Processing Technology*,
Springer Proceedings in Materials 12,
https://doi.org/10.1007/978-981-16-3297-6_6

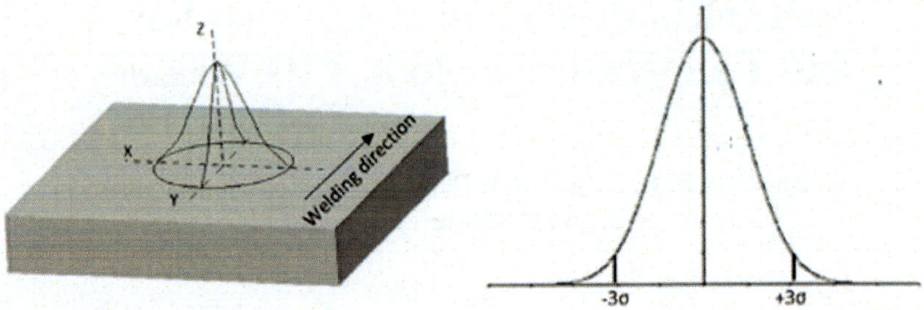

Fig. 1 Gaussian model

2 Temperature Field Distribution

2.1 Gaussian Heat Distribution Model—Ellipsoidal Power Density Distribution Model

Pavelic introduced a Gaussian form of heat distribution in the early study of the temperature field [1]. Pavelic proposed that the heat source should be dispensed and Gaussian flux distribution be deposited on the surface of workpiece. It realizes that this heat source model distributes the heat in an elliptical manner [2]. But this heat source model is helpful for simulation mainly of arc welding in vertically placed manner. This model is widely used to define the energy distribution for arc welding. At the same time, it is inaccurate in modeling an inclined burner [2]. Gaussian heat distribution is using commonly in studies. The pictorial representation of the Gaussian model is given below (see Fig. 1).

Mathematical representation of Gaussian heat distribution is given by

$$q(x, y, z, t) = \frac{\eta U I 6\sqrt{3}}{\pi\sqrt{\pi} abc} exp\left[-3\left(\frac{x}{a}\right)^2 - 3\left(\frac{y}{b}\right)^2 - 3\left(\frac{z + v(\tau - t)}{c}\right)^2\right] \quad (1)$$

where U—Welding voltage, η—Heat source efficiency, I—Welding current, v—Welding speed, a, b, c—Weld pool parameter, and t—Time.

This heat source model is useful for arc welding on a vertically placed burner.

2.2 Goldak Heat Source Model—Double Ellipsoidal Power Density Distribution Model

Afterward, Goldak brought forward a new model using two ellipsoidal heat sources. One defines heat input at the front and another one defines heat input at the rear part

Fig. 2 Goldark model

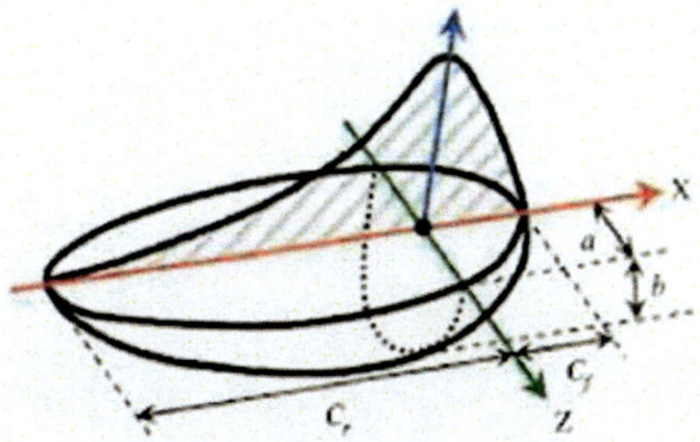

of the ellipsoid, called as a double ellipsoidal model. With the introduction of the double ellipsoidal model, the size and shape of a heat source can be simply changed to model. This heat source provides a good explanation of energy distribution to both sections of the ellipsoid. The mathematical representation heat distribution of this heat source model is given below.

$$q(x, y, z, t) = \frac{f_i \eta U I 6\sqrt{3}}{\pi\sqrt{\pi} abc_i} exp\left[-3\left(\frac{x}{a}\right)^2 - 3\left(\frac{y}{b}\right)^2 - 3\left(\frac{z + v(\tau - t)}{ci}\right)^2\right] \quad (2)$$

where η—Heat source efficiency, *v*—Welding speed, t—Time, f_i—Energy distribution, U—Welding voltage, I—Welding current, c_i—Heat source depth, a, b—Weld pool parameters.

A pictorial presentation of the Goldak heat-source distribution is given in Fig. 2. However, it is not powerful for modeling the oscillating burner.

2.3 *Sabapathy Heat Source Model-Modified Double Ellipsoidal Power Density Distribution*

Sabapathy brings a heat source model after Goldak and it can be used for in-service welding using the wave technique. This method often causes hallow and flatter penetration and distributed more evenly than Gaussian. Sabapathy modified the exponential terms of double ellipsoidal heat source and created his own model, called the Sabapathy heat source model. The mathematical equation for Sabapathy heat distribution is given by

$$q(x, y, z, t) = \frac{fi\eta 6\sqrt{3}}{\pi\sqrt{\pi} abc_i} exp\left[-3\left(\frac{|x|}{a}\right)^{n_1} - 3\left(\frac{|y|}{b}\right)^{n_2} - 3\left(\frac{z + v(\tau - 1)}{ci}\right)^{n_3}\right] \quad (3)$$

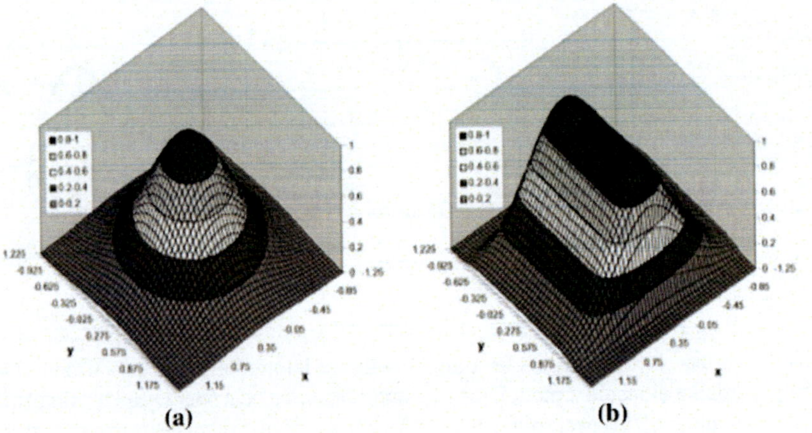

Fig. 3 Sabapathy model

where n_1, n_2, n_3—exponential terms, η—heat source efficiency, v—welding speed, U—welding voltage, I—welding current, a, b—weld pool parameters, t—time, C_i—heat depth and f_i—energy distribution.

The pictorial representation of the Sabapathy model is given in Fig. 3.

Here the various shapes are obtained by changing the power values n_1, n_2, n_3.Sabapathy model is the latest heat distribution model which is devolved to stimulate the low hydrogen arc welding process.

3 Physical Modeling

For the analysis of the effect of welding parameters and geometrical parameters in residual stress, a model of a specimen has to be developed. The standardized dimensions from the literature are taken as length = 100 mm, width = 50 mm, and thickness = 10 mm. As per the given dimension, these two specimens (mild steel and stainless steel) have been modeled and then meshed. This is shown below.

The value of various material properties of mild steel and stainless steel used for this analysis are shown in Tables 1 and 2.

This triangularly meshed block is then analyzed by using thermal and structural parameters.

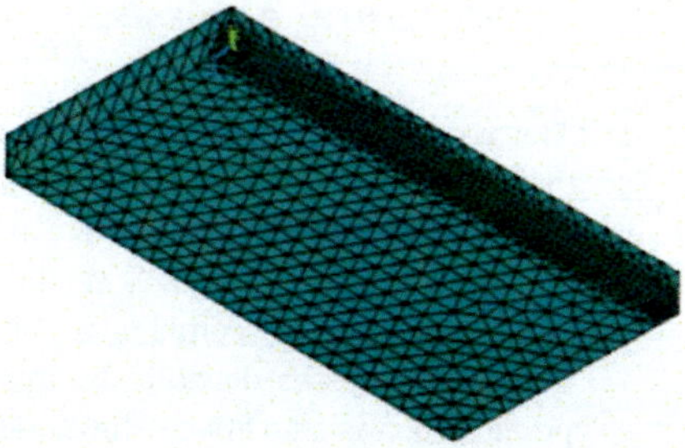

Fig. 4 FEM model

Table 1 Thermal and mechanical properties of mild steel 42

Tensile stress (MPa)	Yield stress (MPa)	Elongation (mm)	Elastic modulus (GPa)	Density (Kg/cm^3)	Co-efficient of thermal conductivity [$\times 10^{-6}/k$]	Thermal conductivity (W/m.k)	Specific heat (J/Kg.k)	Poisons ratio
415	290	20	190	7800	13	51	470	0.29

Table 2 Thermal and mechanical properties of stainless steel sus 304

Tensile stress (MPa)	Yield stress (MPa)	Elongation (mm)	Elastic modulus (GPa)	Density (Kg/cm^3)	Co-efficient of thermal conductivity [$\times 10^{-6}/k$]	Thermal conductivity (W/m.k)	Specific heat (J/Kg.k)	Poisons ratio
515	205	40	193	8000	17.2	16.2	500	0.275

4 Thermal Analysis

Thermal analysis is used to evaluate the temperature distribution across a product based on thermal boundary conditions and specified heat sources. The thermal analysis was conducted by using the FEM model given in Fig. 4. The boundary conditions used for the thermal analysis are given below (see Table 3).

These parameters are selected as per standard conditions, for getting lower residual stress values. Thermal analysis was carried out by using these boundary conditions in the case of both mild steel and stainless steel.

Table 3 Welding parameters

Materials	Current A	Voltage V	Speed mm/s	Weld pool parameters		
				a mm	b mm	c mm
Mild steel	50	12	5	4	3	5
Stainless steel	50	12	5	4	3	5

5 Structural Analysis

The thermal analysis only gives input for structural analysis. Structural analysis gives the exact outcome (effect) of residual stress. Using the same parameters as mentioned in the thermal analysis, structural analysis was carried out and the result obtained from mild steel and stainless steel are given below (see Figs. 5 and 6).

Both tensile and compressive stress show the same nature on longitudinal and transverse stresses on both the materials. Tensile stress is developing at its peak point, which is a positive value and it decreases down. After which the compressive stress also begins from the higher value that starts to compress the material, so that the stress values decreases and come to zero.

The variations in welding parameters and geometric parameters as mentioned above are carried out here.

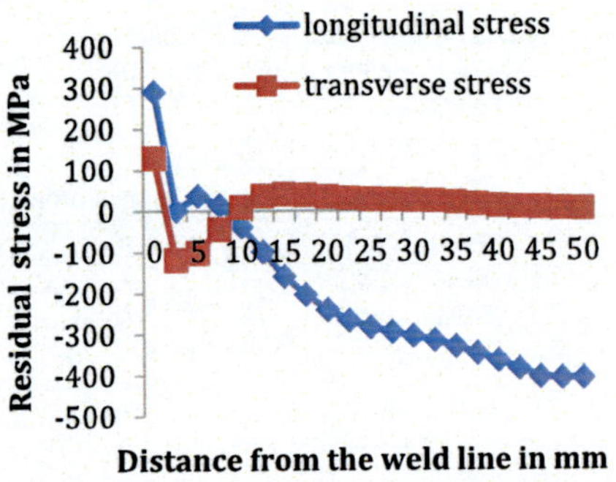

Fig. 5 Longitudinal stress and transverse stress in mild steel

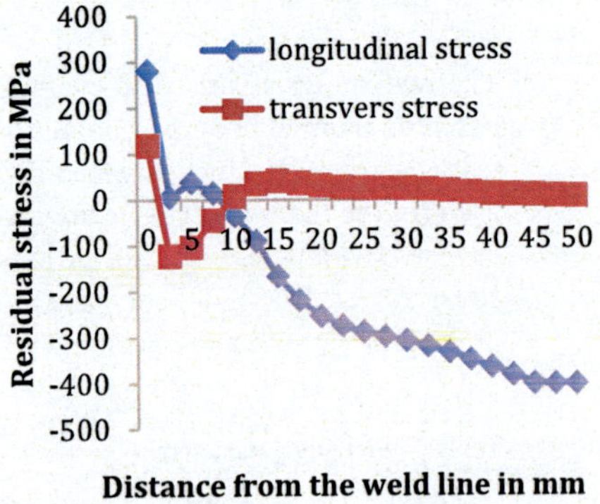

Fig. 6 Longitudinal stress and transverse stress in stainless steel

5.1 *Effect of Welding Parameters in Stress Distribution on Mild Steel and Stainless Steel Welded Joints*

5.1.1 Effect of Weld Current Variation in Residual Stress of Mild Steel and Stainless Steel Joints

The effects generated by current variation in residual stress of mild steel and stainless steel weld joints in longitudinal and transverse directions are given below. Here the current changes to 50 A, 100 A, and 150 A and the remaining parameters are kept the same as above.

By comparing these entire plots with Figs. 5 and 6, the stress value is within the limit. Because the normal 50 A current shows the same stress values as Figs. 5 and 6 and other current shows the variations.

Here both tensile increases and compressive residual stress have no important change while increasing the welding current. But here the stainless steel is better than mild steel due to lower residual stress development.

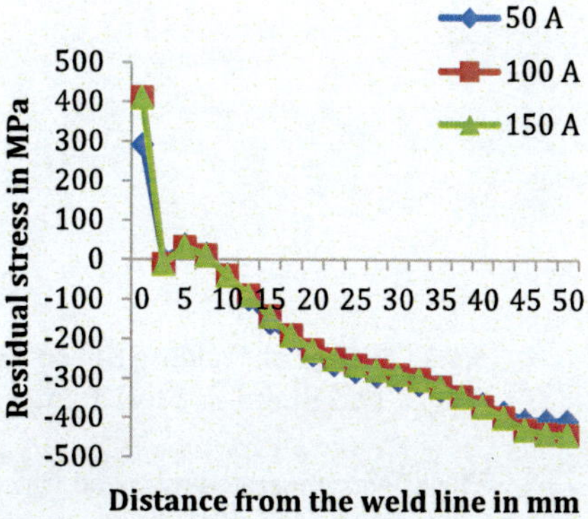

Fig. 7 Effect of current variation on mild steel in longitudinal stresses

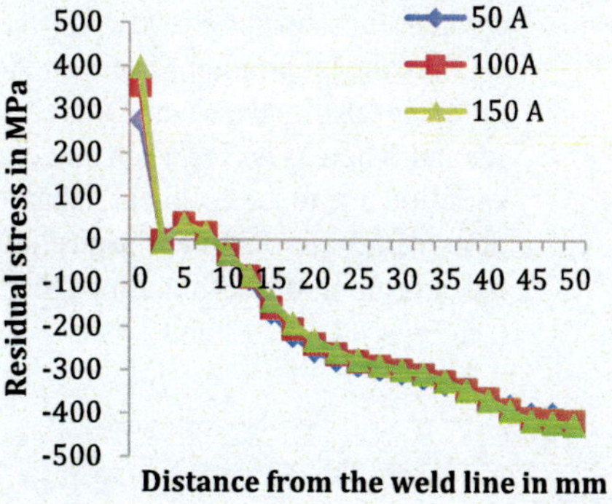

Fig. 8 Effect of current variation on stainless steel in longitudinal stresses

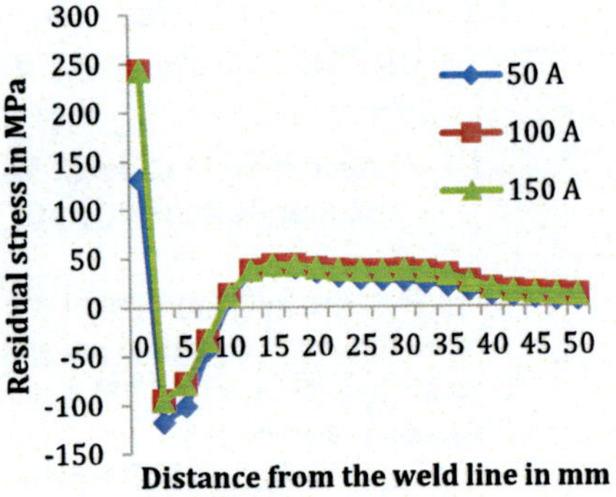

Fig. 9 Effect of current variation on mild steel in transverse stresses

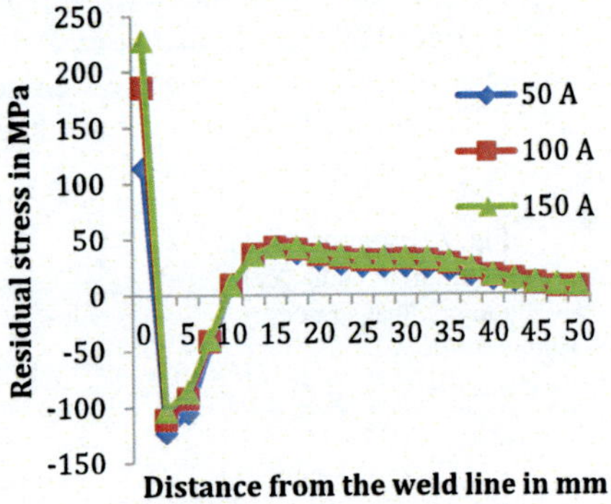

Fig. 10 Effect of current variation on stainless steel in transverse stresses

5.1.2 Effect of Welding Speed Variation in Residual Stress of Mild Steel and Stainless Steel Joints

The effects generated by speed variation in residual stress of mild steel and stainless steel weld joints in longitudinal and transverse directions are shown below. Here the speed changes to 4, 5, 10, 20, and 25 mm per second and the remaining parameters are kept the same as above.

Here the longitudinal stress and transverse stress condition in both materials show the same tensile stress while increasing the speed from 4 mm per second to 25 mm per second. There is no variation that can be obtained. Also, compressive stress shows no variation due to the collinear behavior of all the speeds. So by comparing these entire plots with Figs. 5 and 6, the stress value is within the limit. So it is clear that speed has no role in the formation of residual stress on these materials in either direction (see Figs. 11, 12, 13 and 14).

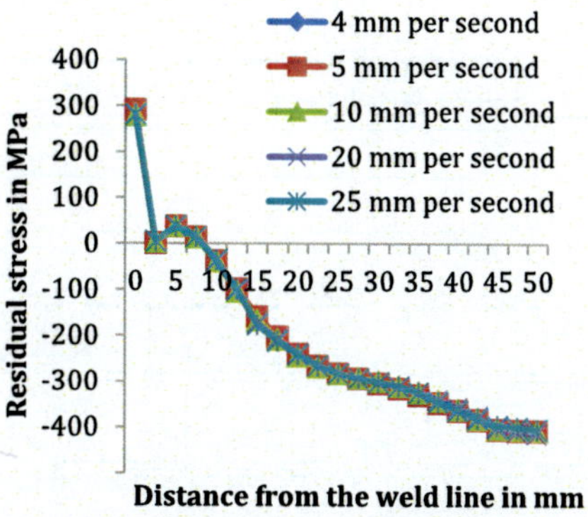

Fig. 11 Effect of speed variation on mild steel in longitudinal stresses

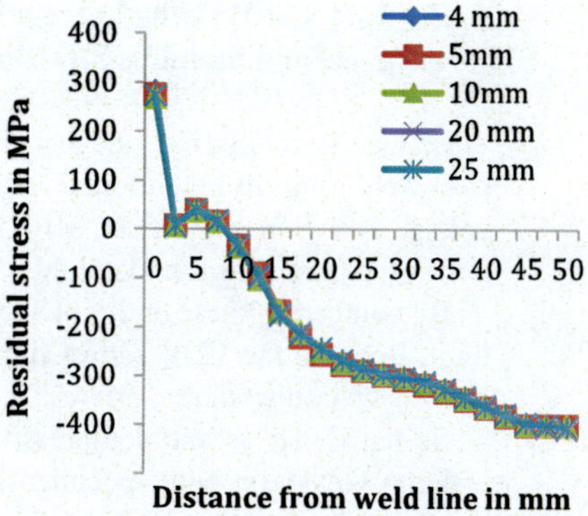

Fig. 12 Effect of speed variation on stainless steel in longitudinal stresses

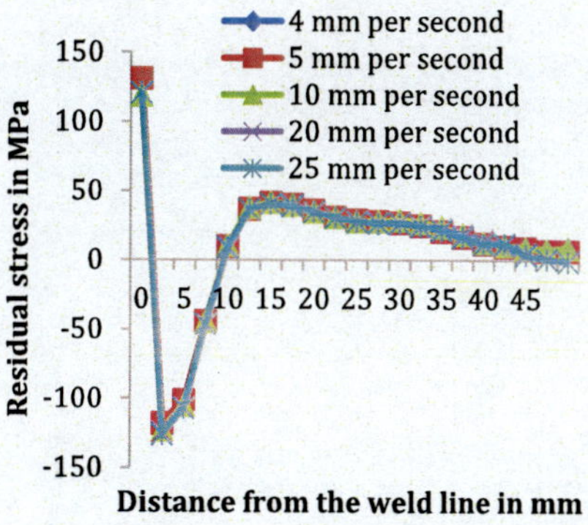

Fig. 13 Effect of speed variation on mild steel in transverse stresses

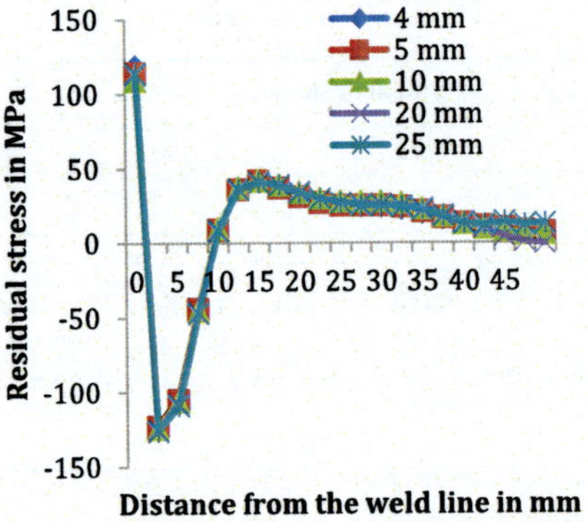

Fig. 14 Effect of speed variation on stainless steel in transverse stresses

5.1.3 Effect of Welding Voltage Variation in Residual Stress of Mild Steel and Stainless Steel Joints

The effects generated by voltage variation in residual stress of mild steel and stainless steel weld joints in longitudinal and transverse directions are given below. Here the voltage variation starts from 10 V and then moves to 12, 15, and 24 V and the remaining parameters are kept the same as above.

By comparing these entire plots with Figs. 5 and 6, the stress value is within the limit. Because the 12 V shows similar variations of Figs. 5 and 6. Other voltages show deviation from it.

So tensile stress and compressive residual stress increase while increasing the welding voltage in both specimens. But here also the stainless steel is better than mild steel due to lower residual stress development.

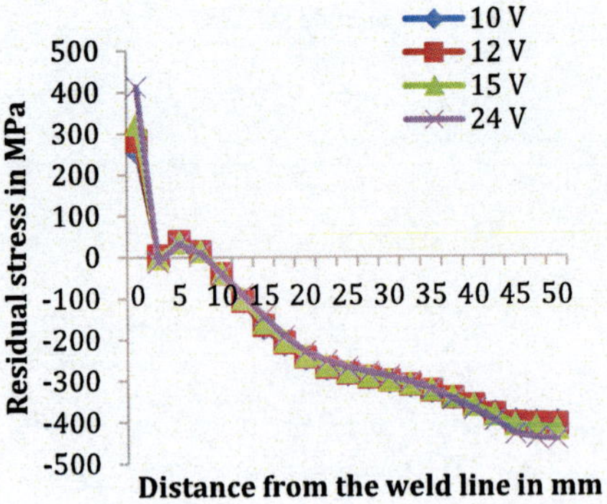

Fig. 15 Effect of voltage variation on mild steel in longitudinal stresses

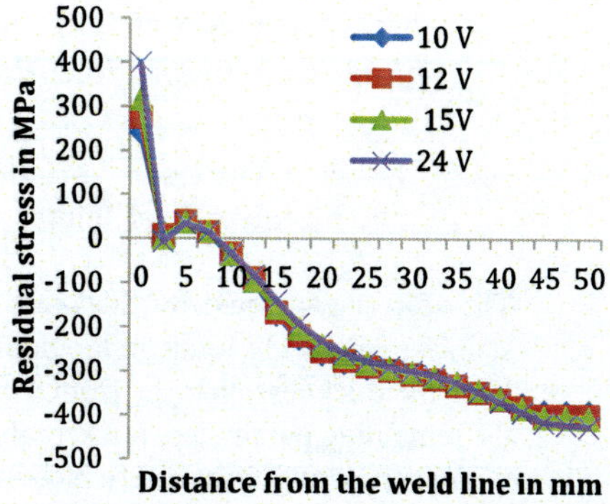

Fig. 16 Effect of voltage variation on stainless steel in longitudinal stresses

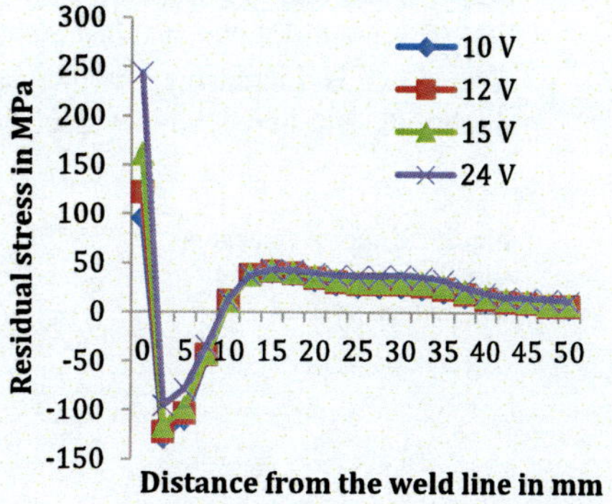

Fig. 17 Effect of voltage variation on mild steel in transverse stresses

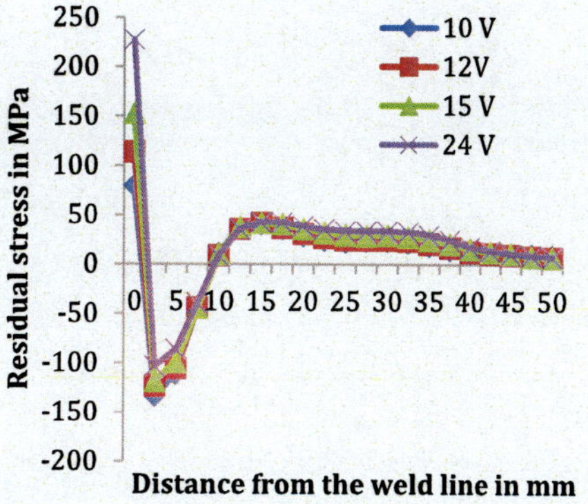

Fig. 18 Effect of voltage variation on stainless steel in transverse stresses

5.2 Effect of Geometrical Parameters in Residual Stress of Mild Steel and Stainless Steel Welded Joints

5.2.1 Effect of Thickness Variation in Residual Stress of Mild Steel and Stainless Steel Joints

The effects generated by thickness variation in residual stress of mild steel and stainless steel weld joints in longitudinal and transverse directions are given below. Here the thickness changes from 4 to 6 mm, then 8 mm, and finally to 10 mm, and the remaining parameters are kept the same as above.

By comparing these entire plots with Figs. 5 and 6, the stress value is within the limit. Because the plate has 10 mm thickness showing similar variations of Figs. 5 and 6. Other thicknesses shows variation in the development of residual stress.

In this condition, the result shows that the tensile stress is fluctuating and compressive stress is increasing by increasing thickness in both the materials in both directions. But here stainless steel shows larger values than mild steel sometimes.

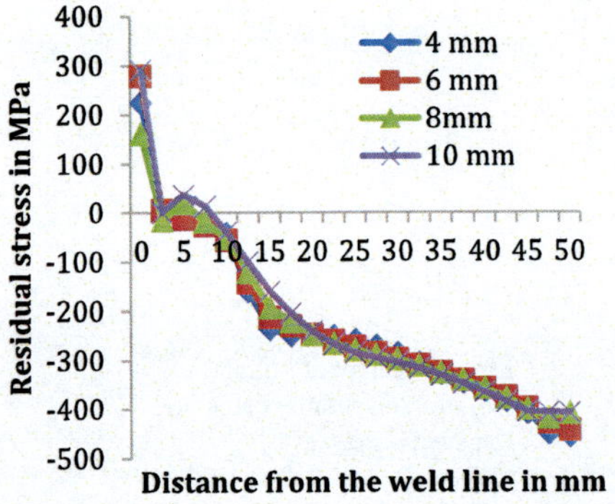

Fig. 19 Effect of thickness variation on mild steel in longitudinal stresses

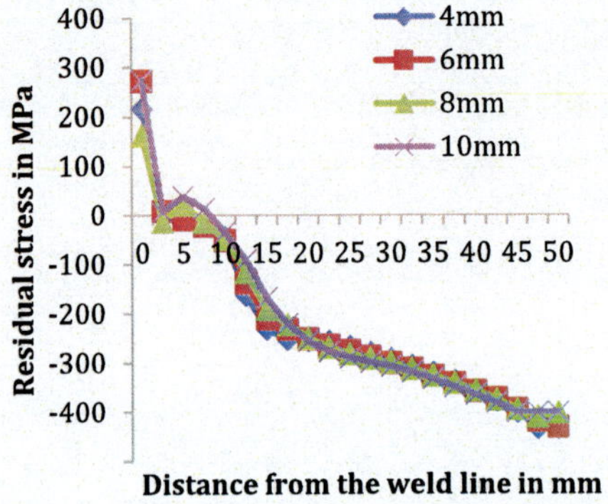

Fig. 20 Effect of thickness variation on stainless steel in longitudinal stresses

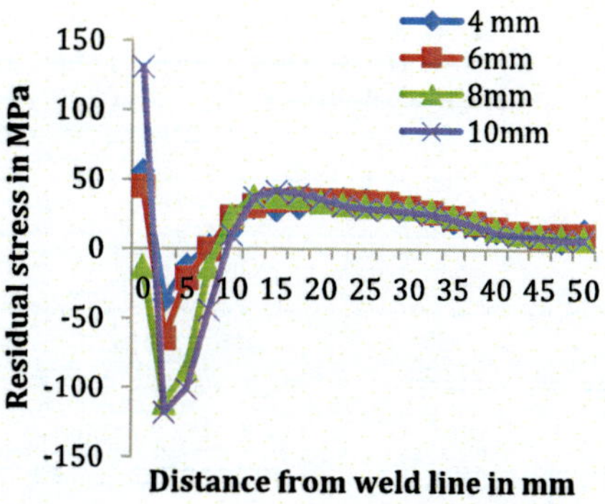

Fig. 21 Effect of thickness variation on mild steel in transverse stresses

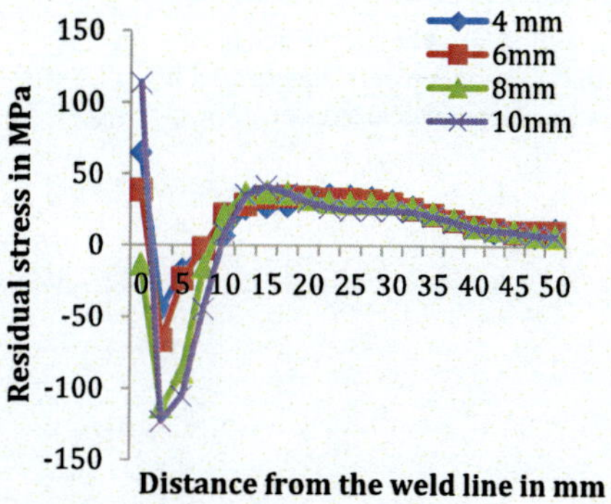

Fig. 22 Effect of thickness variation on stainless steel in transverse stresses

5.2.2 Effect of Height Variation in Residual Stress of Mild Steel and Stainless Steel

The effects generated by height variation in residual stress of mild steel and stainless steel weld joints in longitudinal and transverse directions are given below. Here the height variation is considered from the weld line (weld line is represented as 0 mm). Therefore, the first result obtained from the weld line and then height changes to 3 mm above, 5 mm above, 7 mm above, and finally 10 mm above from the weld line respectively. Still, the remaining parameters are kept the same as above.

By comparing these entire plots with Figs. 5 and 6, the stress values are within the limit. Because weld line showing similar variations of Figs. 5 and 6. Other height variations show deviation from it.

Generally, it is clear that the tensile residual stress and compressive stress were fluctuating on increasing the height. The higher values of stress here also had shown by mild steel.

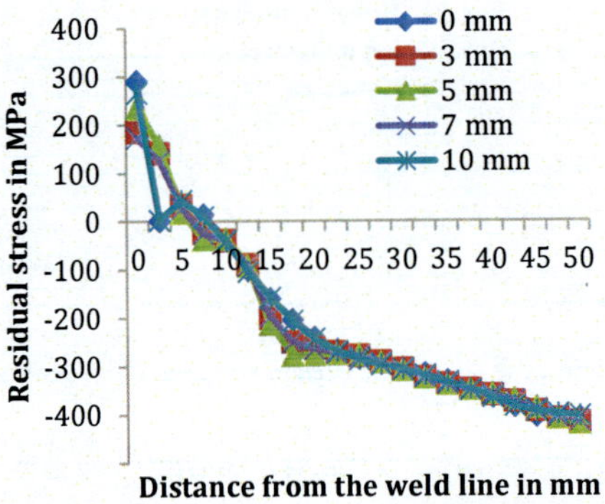

Fig. 23 Effect of height variation on mild steel in longitudinal stresses

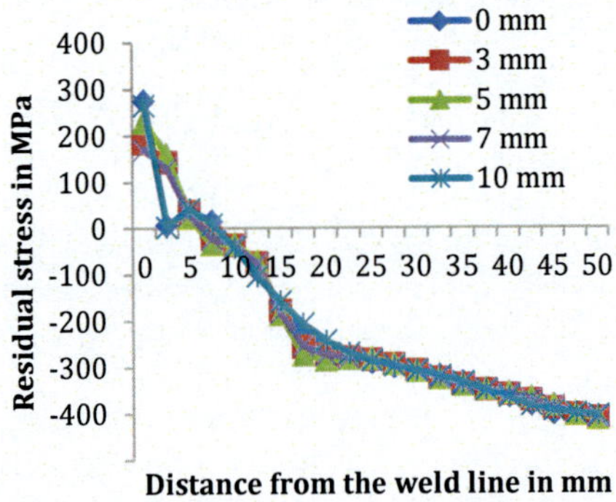

Fig. 24 Effect of height variation on stainless steel in longitudinal stresses

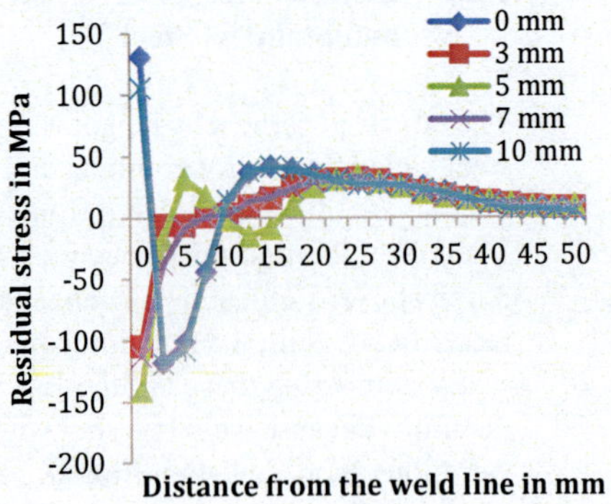

Fig. 25 Effect of height variation on mild steel in transverse stresses

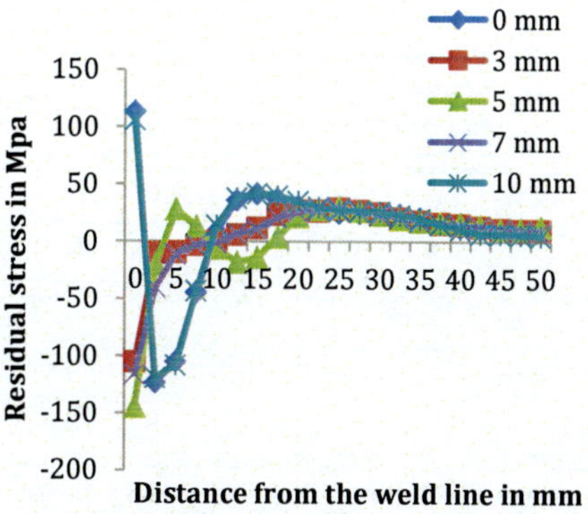

Fig. 26 Effect of height variation on stainless steel in transverse stresses

6 Results and Discussions

Here all these parameter variations in residual stress formation are studied by considering a particular position to validate the result. It is taken as a weld line, because it gives a clear representation of tensile stress compared with other points or positions.

Out of tensile stress and compressive stress, tensile stress shows changes in both directions and materials according to parameter variations. The minimum value of tensile stress gives a perfect weld, which has minimum defect.

The study on tensile stress that shows minimum stress value at weld line is carried out in this chapter.

6.1 Analysis of the Effect of Welding and Geometrical Parameters in Tensile Stress of Mild Steel and Stainless Steel

6.1.1 Analysis of the Effect of Welding Parameters in Tensile Stress of Mild Steel and Stainless Steel

Here the left side of the plot showing the development of residual stress according to current variation. In current variation, mild steel showing higher values than stainless steel in both longitudinal and transverse directions. Out of them, 50 A current giving very less values when compared with other variations. The middle of the plot showing the development of residual stress according to the speed variation. But in the speed variation, stress values are approximately nearer in the case of all longitudinal and transverse directions. By comparing with other variations, it can be negligible. The plot of voltage variation is shown on right side. Here also mild steel showing higher

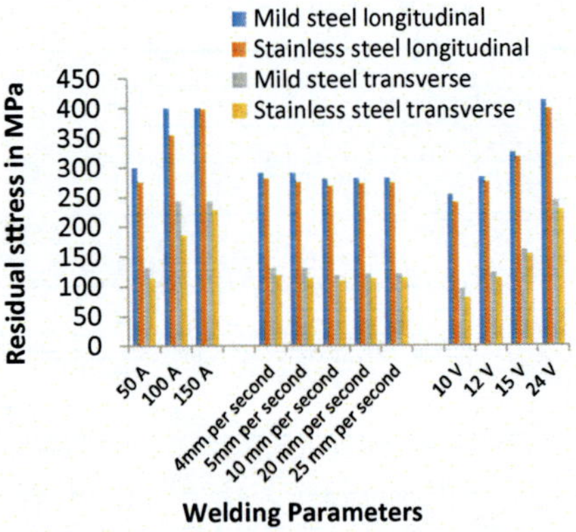

Fig. 27 Analysis of the effect of welding parameters in tensile stress of mild steel and stainless steel

stress than stainless steel in both the directions. The 10 V voltage showing lower stress when compared with other variation (see Fig. 27).

So stainless steel at 50 A current, 5 mm per second speed (standard), 10 V voltage can be chosen to starts this welding process.

6.1.2 Analysis of Effect of Geometrical Parameters in Tensile Stress of Mild Steel and Stainless Steel

Here the left side of the plot shows the development of residual stress according to thickness variation. In thickness variation, mild steel shows higher values than stainless steel in both longitudinal and transverse directions except 4 mm thickness in the transverse direction. Out of them, 8 mm thickness gives very few values when compared with other variations. The plot of height variation is shown on the right side. Here also mild steel shows higher stress than stainless steel in both directions

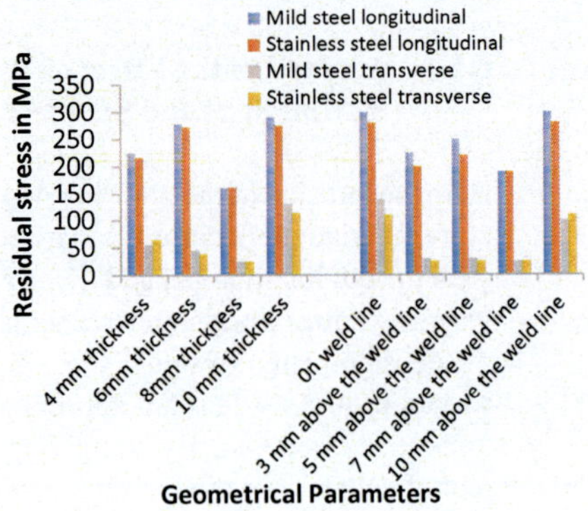

Fig. 28 Analysis of effect of geometrical parameters in tensile stress of mild steel and stainless steel

except for residual stress at top of the specimen. The 7 mm height from the weld line shows lower stress when compared with other variations (see Fig. 28).

So stainless steel having a length of 100 mm, width of 50 mm, and thickness of 8 mm is suitable for this analysis. On this plate, the residual stress obtained is minimum of 7 mm above the weld line.

7 Conclusions

This paper mainly deals with a detailed analysis of residual stress and the development of temperature fields in welding. The main conclusions are the following:

Temperature field and residual stress are the main problems arising during a welding process. FEM has a major role in predicting them. ANSYS is the major tool that can be used for such a numerical analysis.

From thermal analysis, it is clear that stainless steel is producing less temperature than mild steel. But thermal analysis provides only an input to perform structural analysis and the structural analysis gives the development of residual stress as per welding and geometrical parameter.

Welding parameters and geometrical parameters have an influence on the development of residual stress. When considering the welding parameters, voltage has more influence on the development of residual stress compared to other welding parameters. Also, geometrical parameters such as thickness and height have equal importance, but the height variation shows more influence in the development of residual stress.

The study shows that stainless steel can be selected for this analysis than mild steel and the welding parameters such as current, speed and voltage can be selected as 50 A, 5 mm per second, and 10 V, respectively. According to geometrical parameters, the plate should be modeled with 100 mm length and 8 mm thickness. Here the minimum tensile stress is formed 7 mm above the weld line.

This type of analysis is a non-destructive type and can be done without experimenting. So it helps to reduce the cost of work and can be implemented in any type of manufacturing industry.

References

1. Prasad VV, Joy Varghes VM, Suresh MR, Kumar DS (2016) #D simulation of residual stress developed during TIG welding of stainless steel pipes. Procedia Technol 24:364–371
2. Edwin Raja Das (2013) Modelling and prediction of HAZ using finite element and neural network modelling.Advances in Production Engineering and management 8:13–24
3. Joy Varghes VM, Suresh MR, Sivakumar D (2013) Recent developments in modeling of heat transfer during TIG welding-a review. Int J Adv Manuf Technol 64:749–754
4. Goldark J (2005) Computational weld mechanics. Springer, pp 1–35

5. Sabapathy PN, Wahab MA, Painter MJ (2001) Numerical model of in-service welding. Mater Process Technol 118:14–21
6. Kobler D, Tuek J, Taljat B (2004) Finite element modelling of GTA weld surfacing applied to hot work tooling. Comput Mater Sci 31:368–378
7. Grey D, Long H, Maropoulos P (2005) Effect of welding speed, energy input and heat source distribution on temperature variations in butt joint welding. 167:393–401
8. Wu CS, Hu QX, Gao JQ (2009) An adaptive heat source model for finite element analysis of keyhole plasma arc welding. Comput Mater Sci 46:167–172
9. Murphy AB, John JL (2010) Modelling of arc welding: the importance of including the arc plasma in the computational domain. Vacuum 85 (85): 579–584
10. Aksari Mousavi SAA, Miresmaeili R (2008) Experimental and numerical analysis of residual stress distribution in TIG welding process for 304L stainless steel. Mater Process 208:383–394
11. Wu CS, Sun IS (2002) Numerical analysis of temperature field during double sided arc welding of thick materials. Comput Mater Sci 25:457–468
13. Withers PJ, Turski M, Buttle DJ (2008) Recent advances in residual stress measurement. Int J Press Vessel Pip 85:118–127
14. Nima Yazdian, Derakhshan ED, Kovacevic R (2018) Numerical prediction and experimental analysis of residual stress field and generated distortion in hybrid laser welded thick plates of high strength steels. Int J Adv Manuf Technol 18
15. Long GJ (2003) Neutron diffraction 2:83–99

Optimization of Process Parameters for Friction Stir Welding of Aluminium Alloy AA5052-H32 by Using Taguchi Method

Pradyumn Kumar Arya, Neelesh Kumar Jain, and M. Jayaprakash

1 Introduction

In applications where high strength is required compared to the weight ratio and widely used in aircraft structures, marine, fuel tank, metalworking, fuel storage, etc., aluminium alloy AA5052 is also used. For joining of aluminium alloys via fusion welding techniques, various defects have been found like porosity formation, residual stresses, low distortion, hot cracking, etc. [1–3]. In various industries, fusion welding such as metal inert gas (MIG) and tungsten inert gas (TIG) welding has been a more frequently adopted approach for lightweight materials like aluminium alloy, magnesium alloy, etc. [4]. Friction stir welding was another advanced method specifically developed to weld Al and its alloy, which avoids the issues associated with the method of fusion welding [5–7]. Due to the great modifications in the joint strength, hardness, and toughness, FSW could be seen to be a quite flexible and inexpensive strategy for joining Al alloy [8]. Excessive plastic deformation as well as frictional heating facilitate the development of a fine grain and elongated grain microstructure within the stir region and TMAZ region [9]. FSW tool provides substantial heat energy for material movement and conducts defects-free welding with a low FSW table movement as well as high rotational speed of tool [10]. Cavaliere et al. [11] show the influence of FSW variables mostly on microstructure and mechanical characteristics of aluminium alloy AA6082 joint produced by FSW. From a lower speed, the yield strength and ductility increase and begin to reduce by enhancement of table movement afterward. Xu et al. [12] represent the effect on temperature growth, microstructural and mechanical characteristics of FSW thick 2219-O Al alloy

P. K. Arya (✉) · N. K. Jain
Disipline of Mechanical Engineering, Indian Institute of Technology Indore, Indore, MP, India

M. Jayaprakash
Disipline of Metallurgy Engineering and Materials Science, Indian Institute of Technology Indore, Indore, MP, India

R. Prasad et al. (eds.), *Advancement in Materials Processing Technology*,
Springer Proceedings in Materials 12,
https://doi.org/10.1007/978-981-16-3297-6_7

joints were studied. There are highly fine and equiaxed grains in the microstructure in weld stir region but elongated grains found in TMAZ and HAZ. As rotational speed increases, percentage elongation reduces, but weld strength (yield and tensile strength) of joints increase. A larger nugget grain size was shown by the conical shoulder and a smaller grain size was shown in the welds by the scrolled shoulder.

The weld created with the conical shoulder shows an elongation reduction of 30% and the scrolled shoulder shows an elongation reduction of 70%, hardness reduction of 15% with base metal [13]. Materials microstructure emerges with extremely fine and equiaxed grain for FSW AA6056 Al alloy and material hardness reaches higher values under all conditions [14]. With the enhancement of FSW table speed and reducing rotational speed of tool, zigzag pattern and onion ring were observed in weld stir zone and minimum hardness area changes from HAZ to weld stir zone [15, 16]. Fujii et al. [17] examined microstructural and mechanical characteristics of AA1050-H24, AA5083-O, and AA6061-T6 FSW with different tool shapes of threadless columns, thread columns, and triangular prisms. The best mechanical properties developed by the columnar tool for FSW of AA1050-H24 with very low deformation resistance. Tool shape does not affect the microstructures and mechanical properties of the joints for the FSW of AA6061-T6, whose deformation resistance is relatively mild. Tool design need not disturb the microstructural and mechanical characteristics of joints for FSW of AA6061-T6 but FSW of AA5083-O, which has a relatively high resistance to deformation, weldability becomes dramatically impacted by FSW rotational speed. Scialpi et al. [18] analyzed the influence of three tools with different shoulder surfaces like cavity, fillet and scroll, and fillet on microstructures and mechanical characteristics of FSW AA6082-T6. Shoulder surface with fillet and cavity is considered optimal technique in all three different shoulder geometries because it increases the joint longitudinal and traverse power and provides the best surface of the crown. The joint strength of welded aluminium alloys AA6061 with five different profiles of tool pins has been studied and weld developed by square shape tool pin have better tensile properties compared to other joints [19]. Recrystallized grain size throughout the nugget region increases as axial load rises and base material's elongated grain form is transformed into fine recrystallized grains [20]. Dawood et al. [21] analyzed various tool pin geometries, and their effect on microstructural and mechanical characteristics of friction stir welding of aluminium alloy AA6061. Result shows that maximum weld strength and hardness is obtained by the triangular pin shape of tool and the smallest weld strength and microhardness is reported by the square pin shape of tool. In situation of narrow diameter of shoulder and FSW pin shape showing a narrow HAZ, less heat is produced by friction. Adalarasan et al. [22] developed Taguchi's L9 orthogonal set and grey relational observation in order to optimize the welding variables of the TIG welded AA 6061 aluminium alloy. The findings concluded that welding current and gas flow rate contributions had a significant effect on welding efficiency.

2 Experimental Procedure

In the present analysis, the reference material used to be 3 mm thickness, better strength AA5052-H32 aluminium alloy comprising magnesium as its key components. Chemical content and mechanical characteristics of AA5052 aluminium alloy are described in Tables 1 and 2, respectively.

For the current work, aluminium alloy plate AA5052 with dimensions of 100*50*3 mm^3 was being used. Before even the weld was started, the aluminium alloy 5052 sheets had been rubbed by the steel brush and cleaned with acetone. The FSW machine was a conventional vertical milling machine that was converted into an FSW using an FSW setup. The sheets become gripped within the fixture to be welded area frame, and the FSW tool is attached within the set situated inside the vertical shaping machine head. The machine is ready to fix FSW tools on tool holders and plates on fixture and also uses a mix of different tool rotational speeds and FSW table speeds. The photo of the set-up and welding process of the fixture on the workpiece is shown in Fig. 1.

Table 1 AA5052 Al Alloy Chemical composition

Material	Mg	Si	Cu	Cr	Mn	Fe	Zn	Ti	Al
AA6061	2.6	0.11	0.1	0.25	0.5	0.5	0.2	0.2	Bal.

Table 2 AA5052 Al Alloy Mechanical properties

Material	UTS (Mpa)	% elongation	Microhardness $(HV)_{0.1}$
AA6061	228	17	68

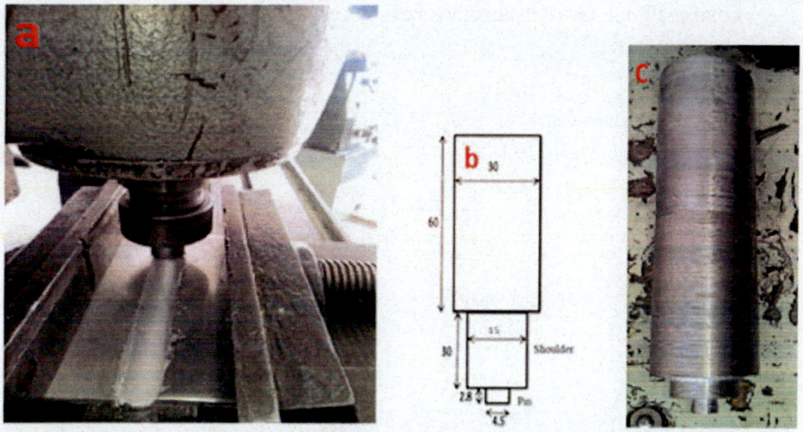

Fig. 1 **a** FSW setup on milling machine, **b** Dimensions of tools, **c** H13 Tool steel for FSW of AA5052

Table 3 Levels and control variables

Control variables	Unit	Level 1	Level 2	Level 3
Rotational speed	RPM	600	1000	1500
Table speed	mm/min	40	60	80
Shoulder diameter	mm	9	12	15

With a simple turning operation on the lathe machine, the FSW tool was prepared from tool steel rod of H13 materials. FSW tool has a shape of a concave shoulder with cylindrical shape pin with a shoulder diameter of 9, 12, 15 mm, diameter of pin 4.5 mm, and length of pin 2.8 mm. Rotational speed of tool, FSW table speed, diameter of shoulder are the parameters defined for current investigation. The preferred process variables and their levels are shown in Table 3. This would be the experimental layout in which the work is carried out.

The Taguchi method is an outstanding factual description of the test instrument needed for evaluating the influence on output variables of process parameters. Orthogonal L9 array can be seen in Table 4. Taguchi includes three kinds of consistency characteristics: small is better, nominal is better, and greater is better. To analyze and predict the correlation between such a response variable and one or maybe more predictor variables, ANOVA is close to regression. To evaluate which parameter significantly affects the output characteristics, statistical software with an ANOVA analytical tool was being used.

The signal to noise (S/N) ratio tests under various noise situations, the response differs relative to the nominal or target value. Based on the needs of your analysis, you could decide from various S/N ratios. Taguchi presented three kinds of performance characteristics small becomes best, nominal becomes best, and greater becomes best. The signal to noise (S/N) proportion becomes computed with each variable level combinations. Greater becomes best formula, S/N proportion through utilizing log of base 10 is:

Table 4 Qrthogonal array of Taguchi L9

Test run	Rotational speed (RPM)	Welding speed (mm/min)	Shoulder diameter (mm)
S1	600	40	9
S2	600	60	12
S3	600	80	15
S4	1000	40	12
S5	1000	60	15
S6	1000	80	9
S7	1500	40	15
S8	1500	60	9
S9	1500	80	12

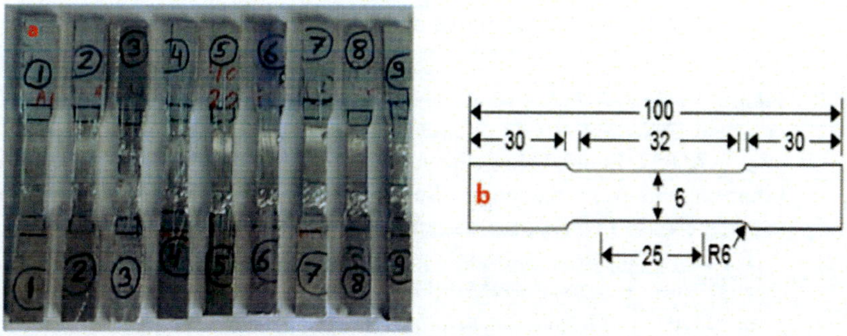

Fig. 2 **a** Tensile test samples **b** graphical view of tensile test sample

$$S/N = -10 * \log (\Sigma (1/Y2)/n)$$

while, Y = responses for mixture of the variable level specified and n = number of responses for mixture of variable level.

According to L9 orthogonal series, FSW was performed on a total of nine samples with completely different welding speeds, shoulder diameter, and tool rpm. After getting prepared FSW specimens, samples of suitable unit area sizes were sliced for different mechanical testing according to ASTM standards. The tensile specimen was developed in accordance with the ASTM E-8 standard and Fig. 2a, b accordingly reflect its graphical representation. The Vickers microhardness evaluation had been investigated at different points over through the cross-section of the FSW regions. Microstructure research had been conducted on specimens to interpret the property when welding and modifications in structure morphology. First, for surface finish, specimen area units extracted from welding sheets and subsequently, metal papers of grit sizes 100, 320, 600, 1000, 2000 have been used. Finally, the samples will be polished with the hifin diamond compound on a single disc polishing system and the etchant (1.5 ml HCL + 1 ml HF + 10 ml HNO3 + 85 ml water) being utilized to examine the microstructure.

3 Result and Discussions

The purpose of such study becomes to define mechanical characteristics (in terms of ultimate tensile strength and hardness) as well as microstructural analysis of AA5052 aluminium alloy friction stir welding.

3.1 Mechanical Properties and Microstructure Analysis

Figure 3 displays the influence of rotational speed of FSW tool, shoulder diameter, and FSW table speed on weld joint strength of the FSW of the aluminium alloy AA5052. Figure below illustrates that owing to excessive plastic flowing and proper combining of aluminium alloys, the joints formed by high tool rotation speed experienced the highest tensile strength.

Because of macroscopic cracks present in the weld region, the joints were formed through lower rotational speed of FSW tool indicates lower weld joint strength and joint efficiency. Microhardness profiles derived from 1000 rpm rotational speed, 12 mm shoulder diameter, and 40 mm/min welding speeds for welded specimens are shown in Fig. 4. Moreover, the advancing side (AS) displays greater hardness values relative to the retreating side (RS), but the highest hardness value was located in the middle zone between the TMAZ and SZ of the AS. In the existing condition, due to the recrystallization and recovery process, the rise in hardness values in the SZ can be caused by the existence of large particle quantities.

The optical microstructure images taken from the various points marks in the macrostructure images and microstructure at various zones are shown in Fig. 5a–d at FSW tool rpm of 1000, FSW table speed of 40 mm/min, and diameter of shoulder 12 mm. Defects including some holes, cracking and surface galling have not been noted inside and near weld stir zone. Including fluctuations of precipitate size as well as its distribution after friction stir welding, there are obvious microstructural modifications. In the HAZ, the precipitates become slightly coarser, but in the nugget region, they become finer.

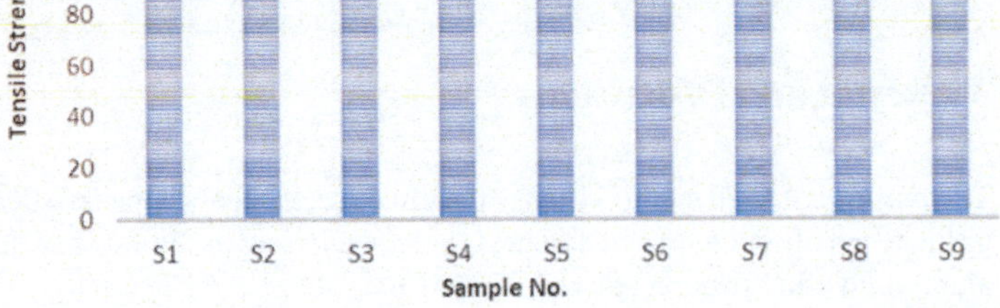

Fig. 3 Shows the tensile strength of joint

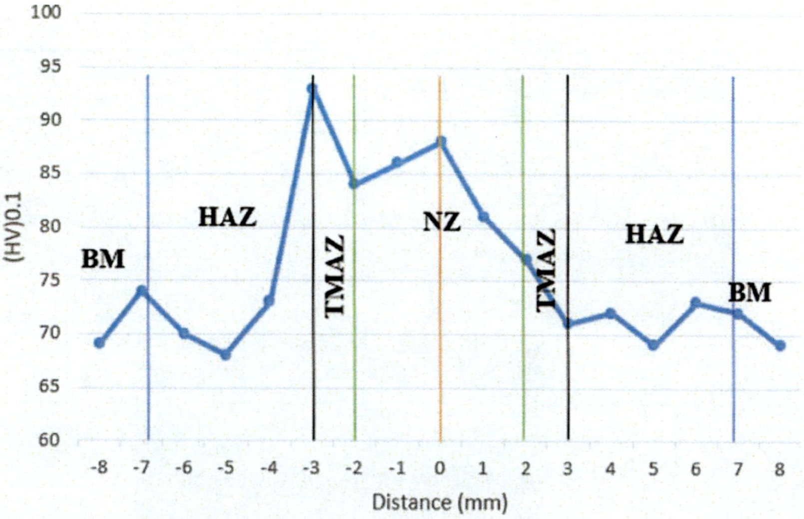

Fig. 4 Shows the hardness distribution of FSW sample

3.2 *Analysis of Tensile Strength*

The signal to noise (S/N) proportion for joint strength becomes estimated using the function of greater become best. As shown in Table 5, the outcome of the S/N ratio for weld joint strength seems to be.

The variance outcome analysis for FSW joint strength can be seen in given Table 6. FSW table speed indicates a greater contribution of 71.84%, diameter of tool shoulder indicates 15.18% contribution, and smallest contribution of 9.66% via rotational speed of FSW tool from the outcome of ANOVA for FSW joints strength. Residual errors were identified as 3.33%.

The major influence of S/N proportion of the FSW joint strength can be seen in Fig. 6. The appropriate FSW method parameter level was estimated through Fig. 6 which are FSW rotational speed of 1000 rpm, FSW table speed of 40 mm/rev, and diameter of shoulder of 12 mm. The average of each response attribute (S/N ratios) with every factor level was shown within response tables. The tables contain ranking that compares the relative magnitude of results, based on Delta statistics. It was concluded through response Table 7 that FSW table speed seems to have the strongest impact on S/N ratio, while the diameter of tool shoulder seems to have the second strongest effect and FSW rotational speed seems to have the lowest impact on S/N ratio.

Fig. 5 Microstructure **a** NZ, **b** TMAZ **c** HAZ **d** HAZ of FSW joint at 40 mm/min, 1000 rp, m and 12 mm

4 Conclusion

The Taguchi method was being used in this study to achieve an ideal condition for Friction Stir Welding of AA5052 aluminium alloy. By using ANOVA, experimental findings were assessed and the following conclusions have been developed.

- To figure out the optimum levels of procedure variables in FSW, Taguchi's experimental methodology model was also used. The optimum FSW tool speed of rotation, FSW table speed, and diameter of FSW tool shoulder levels becomes 1000 rpm, 40 mm/min, and 12 mm, respectively.

Table 5 Calculated S/N ratio of joint strength

Test run	Rotational speed (RPM)	FSW table speed (mm/min)	Diameter of shoulder (mm)	S/N ratio
S1	600	40	9	43.9825
S2	600	60	12	44.2560
S3	600	80	15	42.2394
S4	1000	40	12	45.1112
S5	1000	60	15	44.0970
S6	1000	80	9	43.1140
S7	1500	40	15	44.5921
S8	1500	60	9	43.3599
S9	1500	80	12	43.2495

Table 6 ANOVA result for FSW joint strength

Information	DF	Seq SS	Contribution (%)	Adj SS	Adj MS	F-value	P-value
Rotational speed of tool (RPM)	2	183.53	9.66	183.53	91.76	2.90	0.256
Welding speed (mm/min)	2	1365.32	71.84	1365.32	682.66	21.57	0.044
Shoulder diameter (mm)	2	288.44	15.18	288.44	144.22	4.56	0.180
Error	2	63.30	3.33	63.30	31.65	–	–
Total	8	1900.59	100.00	–	–	–	–

- Welding speed plays a crucial role in this research method and shows more contribution of 71.84%, Shoulder diameter shows 15.18% contribution, and tool rotational speed has the lowest contribution of 9.66%.
- Optimal level of rotational speed of tool 1000 rpm, FSW table speed 40 mm/min, and diameter of tool shoulder 12 mm, shows the maximum tensile strength of 180.12 MPa owing to extreme plastic flowing as well as properly combining of aluminium alloys.
- Advancing side (AS) displays greater hardness values relative to the retreating side (RS), but the maximum hardness result was estimated between TMAZ and NZ of AS in the middle regions. In the HAZ side, the precipitates become slightly coarser, but in the nugget region, they become finer.
- The Taguchi method efficiently optimized the parameters of the friction stir welding operation.

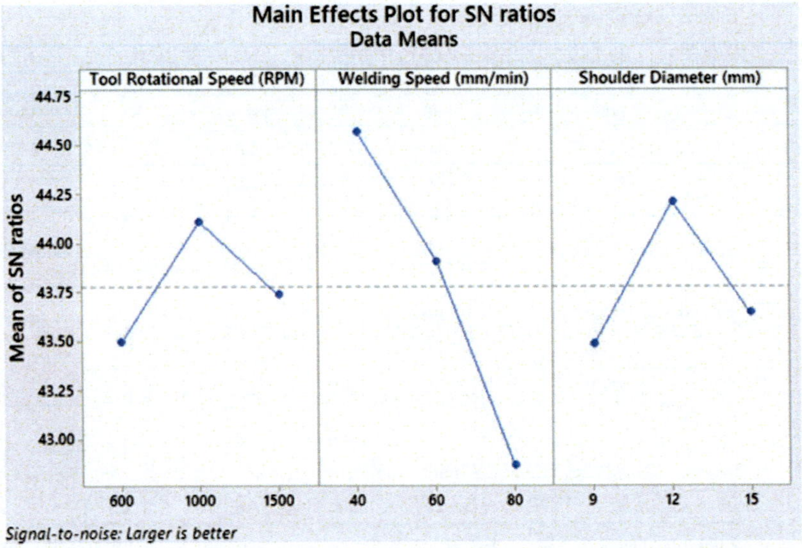

Fig. 6 Major influence graph for S/N ratio

Table 7 Response for S/N ratios

Levels	FSW rotational speed (RPM)	FSW table speed (mm/min)	Diameter of shoulder (mm)
1	43.49	44.56	43.49
2	44.11	43.90	44.21
3	43.73	42.87	43.64
Delta	0.61	1.69	0.72
Rank	3	1	2

References

1. Taban E, Kaluc E (2007) Comparison between microstructure characteristics and joint performance of 5086–H32 aluminium alloy welded by MIG, TIG and friction stir welding processes. Kovove Mater 45(5):241
2. Kah P, Rajan R, Martikainen J, Suoranta R (2015) Investigation of weld defects in friction-stir welding and fusion welding of aluminium alloys. Int J Mech Mater Eng 10(1):26
3. Naik AB, Reddy AC (2018) Optimization of tensile strength in TIG welding using the Taguchi method and analysis of variance (ANOVA). Thermal Sci Eng Prog 1(8):327–339
4. Mathers G (2002) The welding of aluminium and its alloys. Woodhead Publishing, 248 pages. ISBN 1 85573 567 9
5. The Welding Institute TWI. www.twi.co.uk
6. Dawes C, Thomas W (1995) Friction stirs joining of aluminium alloys. TWI Bull 6:124
7. Miles MP, Decker BJ, Nelson TW (2004) Formability and strength of friction-stir-welded aluminium sheets. Metall Mater Trans A 35A:3461–3468

8. Arya PK, Gupta G, Rajput AK (2016) A Review on friction stir welding for aluminium alloy to steel. Int J Sci Eng Res 7(5):119
9. Patel VK, Kumar P, Bhattacharya S (2018) Mechanical, microstructural and sliding wear properties of friction stir welded AA6063-T6 and AA5052-H32 aluminium alloys. Mater Focus 7(1):50–58
10. Kumar N, Patel VK (2020) Effect of SiC/Si_3N_4 micro-reinforcement on mechanical and wear properties of friction stir welded AA6061-T6 aluminium alloy. SN Appl Sci 2(9):1–1
11. Cavaliere P, Squillace A, Panella F (2008) Effect of welding parameters on mechanical and microstructural properties of AA6082 joints produced by friction stir welding. J Mater Proc Technol 200(1–3):364–372
12. Xu W, Liu J, Luan G, Dong C (2009) Temperature evolution, micro-structure and mechanical properties of friction stir welded thick 2219-O Al Alloy joints. Mater Des 30:1886–1893
13. Rodrigues DM, Loureiro A, Leitao C, Leal RM, Chaparro BM, Vilaça P (2009) Influence of friction stir welding parameters on the microstructural and mechanical properties of AA 6016–T4 thin welds. Mater Des 30(6):1913–1921
14. Cavaliere P, Campanile G, Panella FW, Squillace A (2006) Effect of welding parameters on mechanical and microstructural properties of AA6056 joints produced by Friction Stir Welding. J Mater Process Technol 180:263–270
15. Sharma C, Dwivedi DK, Kumar P (2012) Effect of welding parameters on microstructure and mechanical properties of friction stir welded joints of AA7039 aluminium alloy. Mater Des (1980–2015) 36:379–90
16. Bahuguna S, Arya PK, Patel VK (2020) Mechanical and abrasive wear properties of friction stir welded joints of aluminium alloy AA6061-T6 with/without nickel coating. Strojnícky časopis-J Mech Eng 70(2):21–36
17. Fujii H, Cui L, Maeda M, Nogi K (2006) Effect of tool shape on mechanical properties and microstructure of friction stir welded aluminium alloys. Mater Sci Eng A 419(1–2):25–31
18. Scialpi A, De Filippis LA, Cavaliere P (2007) Influence of shoulder geometry on microstructure and mechanical properties of friction stir welded 6082 aluminium alloy. Mater Des 28(4):1124–1129
19. Elangovan K, Balasubramanian V, Babu S (2009) Predicting tensile strength of friction stir welded AA6061 aluminium alloy joints by a mathematical model. Mater Des 30(1):188–193
20. Kumar K, Kailas SV (2008) On the role of axial load and the effect of interface position on the tensile strength of a friction stir welded aluminium alloy. Mater Des 29(4):791–797
21. Dawood HI, Mohammed KS, Rahmat A, Uday MB (2015) Effect of small tool pin profiles on microstructures and mechanical properties of 6061 aluminium alloy by friction stir welding. Trans Nonferrous Metals Soc China 25(9):2856–2865
22. Adalarasan R, Santhanakumar M (2015) Parameter design in fusion welding of AA 6061 aluminium alloy using desirability grey relational analysis (DGRA) method. J Inst Engineers (India) Series C 96(1):57–63

Computational Study of Melting and Solidification Behavior of PCM Thermal Energy Storage System Using Extended Surface

Pragyan Priyadarsini, Asutosh Beuria, Prem Prasad Satapathy, and Sudhansu S. Sahoo

1 Introduction

The solar energy can be considered as a futuristic source of energy, which has the potential to effectively replace the use of fossil fuels in the near future. The smooth use of the solar energy, however, is affected by the daylight hours. Thermal energy storage (TES) systems using a phase change material (PCM) can be used to provide a steady power during the cloudy hours and night hours. Use of TES makes an equilibrium between the energy requirement and supply. The TES system includes storage methods either in the form of sensible heat or latent heat. The temperature of the system increases without phase change in case of sensible type of thermal storage and phase change occurs in latent heat type of storage. Phase change materials are used in latent heat type of storage system. Usually, PCMs have poor thermal conductivity so that charging/discharging rate is slow and hence, a prolonged phase changing process is common in latent heat type storage systems. It affects the storage and recovery of energy so we need to work to increase the heat transfer.

Ibrahim et al. [1] have stated review on combined technology which improves the heat transfer rate and the research gap regarding those. Improvement in heat transfer of PCM was done by many researchers using different fin configurations with experimentation and numerical approaches. Heat transfer enhancement techniques

P. Priyadarsini (✉) · S. S. Sahoo
Department of Mechanical Engineering, Formerly College of Engineering and Technology, Odisha University of Technology and Research, Bhubaneswar, Odisha, India

A. Beuria
Department of Mechanical Engineering, Indian Institute of Technology, Kharagpur, West Bengal, India

P. P. Satapathy
Department of Mechanical Engineering, ITER, Siksha O Anusandhan University, Bhubaneswar, Odisha, India

R. Prasad et al. (eds.), *Advancement in Materials Processing Technology*,
Springer Proceedings in Materials 12,
https://doi.org/10.1007/978-981-16-3297-6_8

have been applied with eight longitudinal fins and eight triangular fins [2]. Erythritol, being a medium temperature PCM was used for the experiment and it was found that the longitudinal fins, with an increased thermal response have a better performance during the charging process [3].

Two-dimensional numerical model using internal and external fins for the charging and discharging behavior with an involvement of pure conduction and natural convection, the effect of number of fins, fin geometry, Stefan Number, PCM material, and the heat exchanger material have been investigated [4]. A few studies are found in literature related to fin parameters like fin spacing, fin diameter, and flow parameters of HTF in shell and tube latent heat storage systems. Buoyancy-driven convection as heat transfer mode during charging of PCM inside a shell and tube heat exchanger has been studied [5]. Considering different techniques like the use of NanoPCM and fins how affects the solidification rate of PCM have been studied by Mahadi et al. [6]. They studied using different cases like NanoPCM with fin combination, alone fin, and alone NanoPCM.

2 Mathematical Modeling

2.1 Computational Domain

In the present investigation, five different geometric configurations (Fig. 1) have been considered and studied. A triplex tube heat exchanger (TTHX) is used in latent heat thermal energy storage where the outer tube and inner tube are heated by flowing heat transfer fluid and the middle tube contains phase changing material in this work. The PCM gets melted by getting heat from both of inner and outer tubes. The model is designed in computational fluid dynamics approach with the help of ANSYS Fluent 16.0. The Cartesian coordinate system (x, y) is chosen for geometry.

The five-physical configuration of the TTHX model considered are (1) PCM having no fins (2) PCM-triangular fins having four configurations. The dimensions of tubes used TTHX have been considered from [2]. Two-sided heating method was considered for charging purposes, simultaneously heat was supplied from both inner and outer tubes during the charging process. PCM surface was at room temperature 27 °C and the calculated minimum temperature for LHTES system was approximately 90 °C.

2.2 Meshing

The computational domain consists of both solid and fluid bodies. Fine meshing is done by using face meshing in sizing tools which are in two-dimensional geometry. The curvature is on and relevance center is fine. Smoothing is taken as medium.

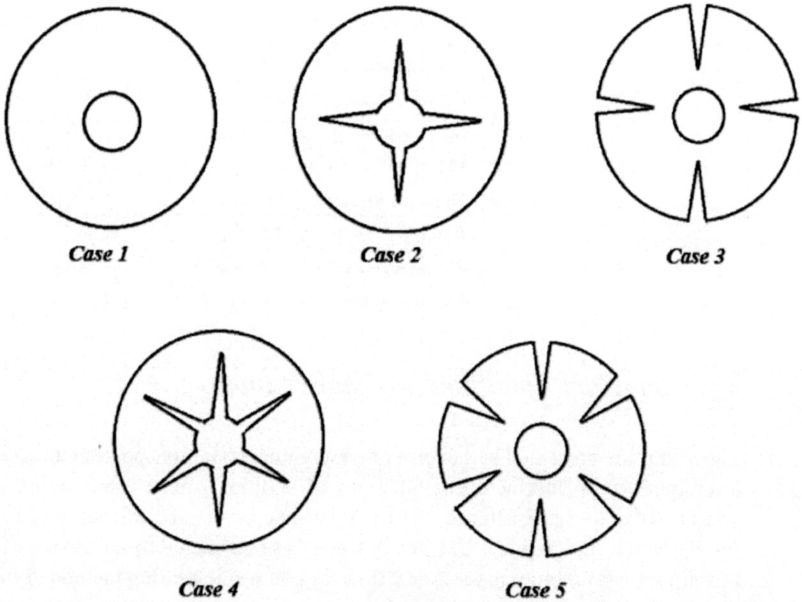

Fig. 1 Case(1)-no fin, Case(2)-Internal four fins, Case(3)-External four fins, Case(4)-Internal six0 fins, Case (5)-External six fins

2.3 Assumptions

For numerical simulation using ANSYS.16, a few assumptions have been adopted.

- The charging process follows Newtonian and incompressible flow.
- Laminar flow and the viscous dissipation effects are primarily negligible.
- Transient method has been considered for analysis and the PCM properties are temperature dependent.
- Both conduction and convection as mode of heat transfer.
- The variation in volume during phase change is negligible and the PCM solid mass is always fixed to the cold walls.

2.4 Material Properties

See Table 1.

Table 1 List of material properties as used in [2]

Properties	PCM (RT 82)
Density, solid ρ_s(kg/m^3)	950
Density, liquid ρ_l(kg/m^3)	770
Specific heat c_{pl}, c_{ps} (J/kg-K)	2000
Latent heat of fusion L (J/kg)	176,000
Dynamic viscosity, μ (kg/m s)	0.03499
Melting temperature, T_m (k)	350.15–358.15
Thermal expansion coefficient (1/k)	0.001
Thermal conductivity, (W/m K)	0.2

2.5 Boundary Conditions and Solver Inputs

The results are presented in the form of temperature contours, mass fraction contours, and liquid–solid interface. For all cases of solidification, surface temperature is constant overall that is 363 K and the inner and outer tube temperature is 300 K. Similarly, for charging of PCM, the process was considered to start when PCM was in solid state (at room temperature). By using both side heating method at inner and outer tubes using temperature (=363 K).

Transient pressure-based solver has been used. Solidification and melting model with laminar, gravity enabled has been applied for pressure velocity coupling simple scheme is used. For discretization, second-order upwind momentum and energy equation, least squares cell-based and presto method for pressure has been used.

2.6 Grid Independence Test

A grid Independence study has been carried out during the course of the present work to find the favorable grid size for the study for the phase change material. The grid independence test is obtained for the un-finned domain. The grid size taken into consideration, are 50,050, 53,264, 57,730, 62,250, 67,268, respectively. The Liquidification time of the PCM (RT-82) is plotted against the grid size (Fig. 2). With rise in grid size, the charging time has been decreased. Also, after a grid size of 57,730, the melting time is observed to have a negligible change. Hence, the grid size in all of the cases under study has been taken well above 57,000 (see Fig. 3).

2.7 Validation

Recent numerical analysis has been done with horizontal TTHX with fins embodying phase change material presented in [2]. It is found that there is a close match between

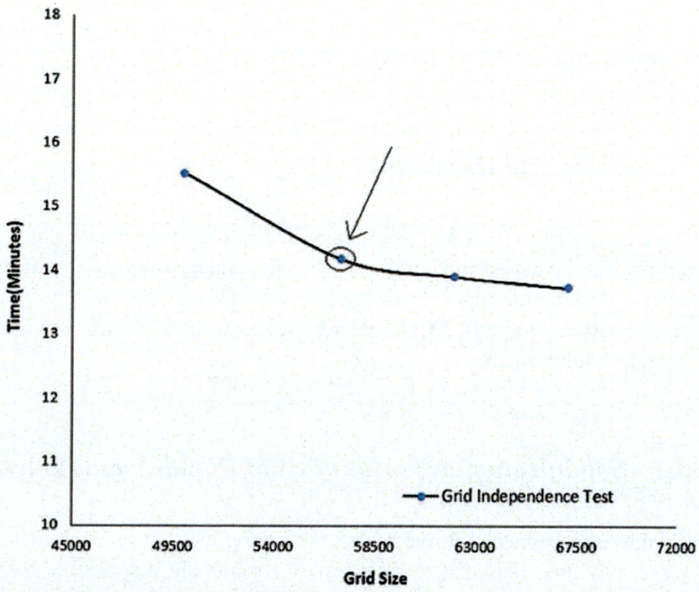

Fig. 2 Grid independence test

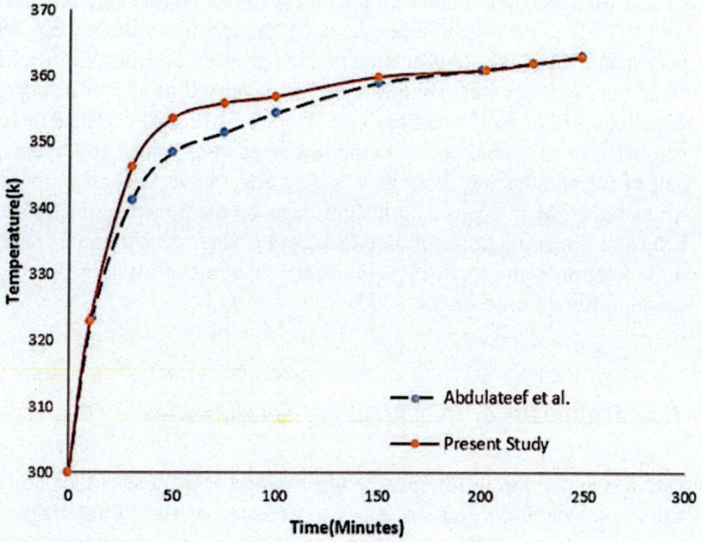

Fig. 3 Validation with [2]

the present result and the results mentioned in literature. So, parametric studies can be done using the validated result.

3 Results and Discussion

Implementing fins to PCM-based geometry leads better enhancement as it is available simply, easy to install and it brings reduction in manufacturing cost Mat et al.[7]. The effects of triangular fins with different fin configurations are compared to without fin configuration geometry. PCM melting time as well as solidification time were also analyzed numerically.

3.1 Solidification Behavior of PCM (Solid–Liquid Interface)

Comparison between all five cases considering melting rate and melting time with respect to time lapsed have been done. As fins are responsible for heat transfer enhancement, finned geometry behaves efficiently than non-fin cases. Liquid PCM gets rapidly solidified near the wall of HTF domain at the initial stage within 0.5 h. Concentric layers are found near the tube walls for no fin geometry and near the tube wall as well as fin void in other cases. With increase in time, these solidified areas get expanded and the solidification process progresses. But solidification slows after sometime. No fin case(1) shows uniformity as well as slower discharging whereas finned cases like case(2), case(3), case(4), case(5), liquid PCM near fins gets squeezed converted to solidified zone. At the last stage after 4 h of solidification, the upper half of the annulus has more unsolidified area due to lower density of liquid PCM than solid PCM. As before in the 2nd stage, all the finned geometry shows the same behavior. Some minor unsolidified area of PCM remains the same near the fins. The mass fraction contours in Fig. 4 indicates how much mass of PCM gets solidified during different time stages.

3.2 Isotherms of PCM During Solidification

Due to temperature difference between liquid PCM and colder heat transfer tube wall, a temperature gradient has been created at the initial stage. Hence buoyancy force is the main motivation for liquefaction at initial stages. Free convection is the heat transfer process initially. With increase in time natural convection is converted into conduction because of increase in solidified area. Domain temperature decreases from HTF temperature to nearly ambient temperature. At initial stage of solidification, nearly 30% of PCM volume has been solidified.

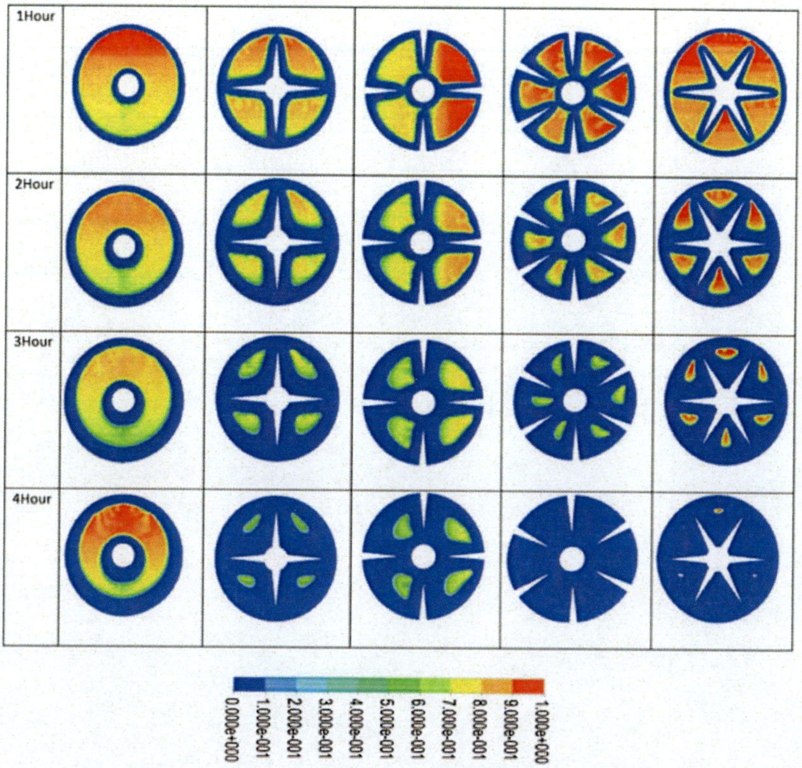

Fig. 4 Mass fraction contours at different time steps for solidification of PCM

Due to greater buoyancy force than gravity, recirculation cells are available in between fins and tube walls. After 1 h of discharging, recirculation cells get shrinked due to dominance of conduction over convection. All the temperature variations for each hour of solidification for all cases have been depicted in Fig. 6. From the figure, it is observed that case(3) has taken maximum time to reach the minimum temperature whereas the external six fins case(5) the least time. The minimum solidification temperature is limited to nearly ambient temperature (see Fig. 5).

3.3 Melting Behavior of PCM (Solid–Liquid Interface)

During the charging process PCM gets melted having temperature greater than its liquidus temperature. Initially, pcm surface is at ambient temperature that is at 300 k. Both side heating of 363 k temperature is applied at the inner and outer wall. Case(1) geometry has been considered from Al-Abidi et al. [4] and compared with our other

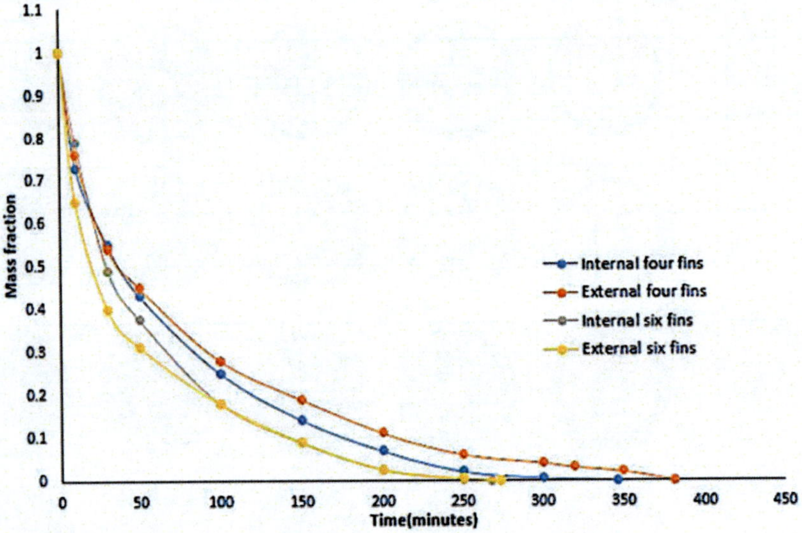

Fig. 5 Mass fraction curve for solidification

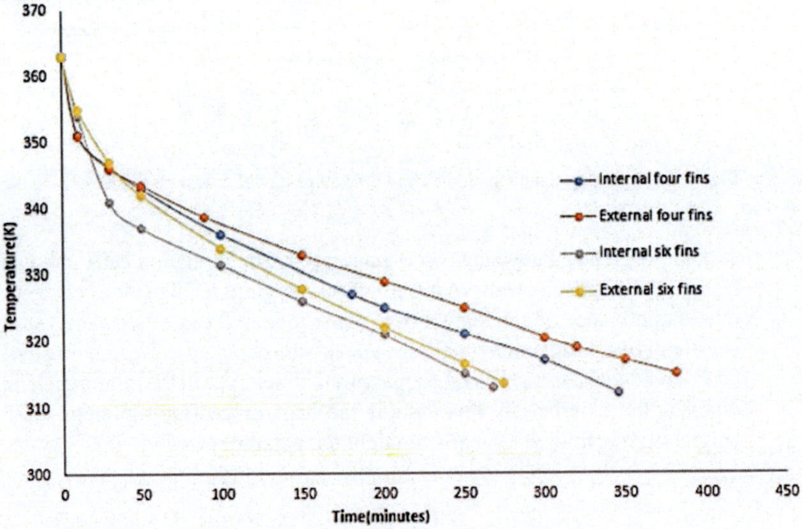

Fig. 6 Temperature curve for solidification

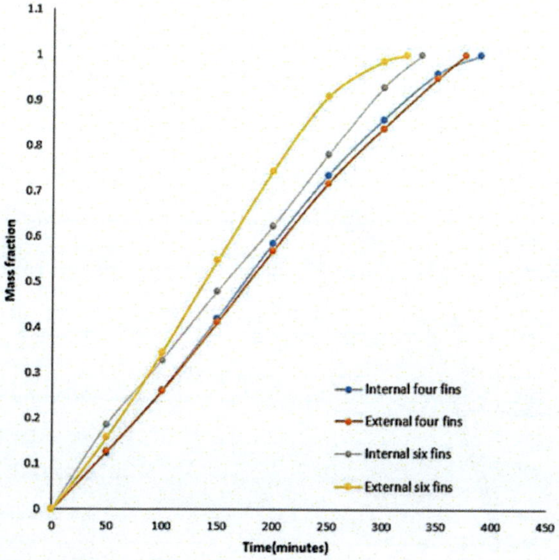

Fig. 7 Mass fraction curve for melting

finned geometries. Heat transfer is done via convection due to two different surfaces; one is solid pcm and fluid flowing hot tubes. When HTF is in contact with solid PCM, some liquid layer will be created all over the annulus. From Fig. 7, it has been observed that melting phenomena before one hour is very insignificant with lesser percentage. For all the fin cases melting occurs near around all the fins both at inner and outer surfaces (see Fig. 8).

Melting process proceeds downward due to gravity effect creating larger recirculation cell after 1 h. Due to prominence in free convection, an increase in heat-transfer rate to the PCM is observed. All the fin geometries provide enhancement in heat transfer. But as the time passes, the circular layer increases and expands to all of the annulus. The melting rates of all cases have been compared to the no fin case(1) Due to more fin areas, case(5) shows the efficient melting than others. Case(2) and case(3) shows few melting time difference and case(4) and case(5) also behaves the same. The total melting timing of PCM for fin has been analyzed. From Graph 8, it is observed temperature rise is linear for all cases before 1 h and the temperature rises in a nonlinear manner after that. After 2 h, the temperature profile leads to a constant profile. Both case(2) and case(3) show same temperature variation and case(4) and case(5) also do the same. The maximum temperature is reached earlier than four fin cases than six fin cases.

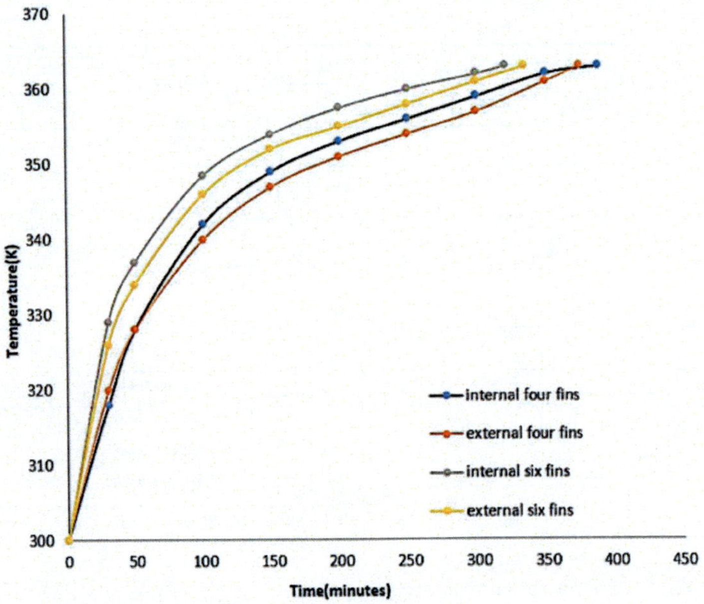

Fig. 8 Temperature curve for melting

3.4 *Isotherms of PCM*

PCM surface temperature rises with course of time from atmospheric temperature to heat transfer fluid temperature. Uniform temperature profile is obtained during initial hours for all cases.

Due to convection cells, temperature rises to maximum temperature as seen in Fig. 9. Case(4) and case(5) show a hike in temperature in comparison to case(1) and case(2). Initially, isotherms are not well spread over the circular area of tubes. Conduction causes a thin layer of liquid PCM during initial hours. In the intermediate phase, the solid–liquid interface grows forward surrounding the two tubes and the buoyancy force leads to natural convection. Solid PCM having higher density has been shrinked to the bottom of the tube. The change in temperature can be observed in Fig. 9. The colder regions are blue in color and the high-temperature zones are in red.

4 Conclusions

Melting and solidification behavior of a commercialized grade phase-changing material, Rubitherm-82 used in horizontal triplex tube heat exchanger thermal

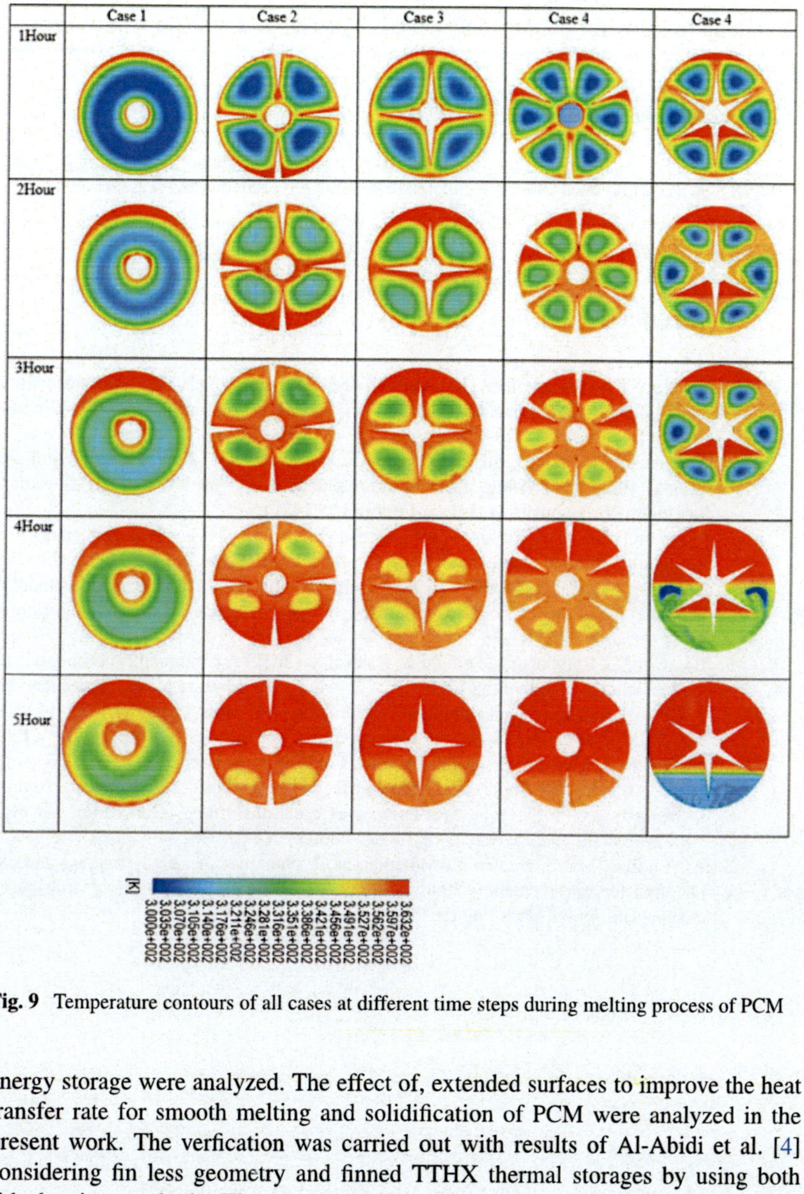

Fig. 9 Temperature contours of all cases at different time steps during melting process of PCM

energy storage were analyzed. The effect of, extended surfaces to improve the heat transfer rate for smooth melting and solidification of PCM were analyzed in the present work. The verfication was carried out with results of Al-Abidi et al. [4] considering fin less geometry and finned TTHX thermal storages by using both side heating methods. The numerical approach used in this study showed that the latent heat thermal energy storage system with external six fin case is efficient than other cases for charging process, while internal six fin case gives the best performance for solidification. More research can be done to improve heat transfer enhancement of phase change materials. The present work will definitely help to study further and develop an effective kind of PCM based thermal storage system.

Acknowledgements Corresponding author who is working as Research Fellow (Dr. Sudhansu S. Sahoo as PI), thankfully acknowledge the financial support from State Council on Science & Technology, Science & Technology Department, and Government of Odisha under research grant ST-SCST-MISC-0061/2018 2682/ dated 24-06-2019 to carry out this present work.

References

1. Ibrahim NI, Al-Sulaiman FA, Rahman S, Yilbas BS, Sahin AZ (2017) Heat transfer enhancement of phase change materials for thermal energy storage applications: A critical review. Renew Sustain Energy Rev 74:26–50
2. Abdulateef AM, Mat S, Sopian K, Abdulateef J, Gitan AA (2017) Experimental and computational study of melting phase-change material in a triplex tube heat exchanger with longitudinal/triangular fins. J Solar Energy 155:142–153
3. Agyenim F, Eames P, Smyth M (2009) A comparison of heat transfer enhancement in a medium temperature thermal energy storage heat exchanger using fins. J Solar Energy 83:1509–1520
4. Al-Abidi AA, Mat S, Sopian K, Sulaiman MY, Mohammad AT (2013) Experimental study of PCM melting in triplex tube thermal energy storage for liquid desiccant air conditioning system. J Energy Build 60:270–279
5. Hosseini MJ, Ranjbar AA, Sedighi K, Rahimi M (2012) A combined experimental and computational study on the melting behavior of a medium temperature phase change storage material inside shell and tube heat exchanger. Int Commun Heat Mass Transf 39:1416–1424
6. Mahdi JM, Nsofor EC (2018) Solidification enhancement of PCM in a triple x tube thermal energy storage system with Nano Particles and Fins. Appl Energy 211:975–986
7. Mat S, Al-Abidi AA, Sopian K, Sulaiman MY, Mohammad AT (2013) Enhance heat transfer for PCM melting in triplex tube with internal–external fins. Energy Conversion and Management 74: 223–236. https://doi.org/10.1016/j.enconman.2013.05.003
8. Al-Abidi A, Mat S, Sopian K, Sulaiman M, Mohammad A (2013) Internal and external fin heat transfer enhancement technique for latent heat thermal energy storage in triplex tube heat exchangers. Appl Therm Eng 53:147–156

Design, Modelling, Fabrication, and Testing of Vertical Milling Machine Fixture for Friction Stir Welding Operation

Sufian Raja, Mohd Bilal Naim Shaikh, Mobin Majeed, Ayush Varshney, and Abdul Samad

1 Introduction

The friction stir welding process, developed and patented by The Welding Institute (TWI, 1991), has many advantages over the other conventional fusion joining techniques [1]. Presently, this technique has significant commercial applications in shipping and marine industries, secondary structures, high speed automotive and railway industries, robotics, aviation industries, electrical distribution assemblies, defense industries, electronic packages, and many others [2, 3]. Furthermore, it is regarded as a cost-saving process due to the following: eliminating consumable costs, equipment flexibility, and reduced manufacturing time, automated nature, and considerable reduction in repair and rework. Therefore, it is considered the most significant development in material joining [4, 5]. High facility and technology cost associated with FSW machines has limited its take up in small-scale industries [6, 7]. Further, purpose-built FSW machines have their manufacturing limitations. Hence, an alternative approach of utilizing capable milling machines for FSW is being exploited by small businesses.

For the utilization of vertical milling machine in performing friction stir welding, there is a need for a well-designed fixture with the following features; hold and position workpiece precisely with respect of rotating tool, withstand at high forces (lateral, transverse, and downward) and temperatures, uniform clamping, facilitate

S. Raja
Faculty of Engineering, Department of Mechanical Engineering, University of Malaya, Kuala Lumpur, Malaysia

M. B. N. Shaikh (✉) · A. Varshney · A. Samad
Department of Mechanical Engineering, Aligarh Muslim University, Aligarh, India
e-mail: mohammed_bilalnaimshaikh@zhcet.ac.in

M. Majeed
College of Science and Engineering, James Cook University, Townsville, Australia

R. Prasad et al. (eds.), *Advancement in Materials Processing Technology*,
Springer Proceedings in Materials 12,
https://doi.org/10.1007/978-981-16-3297-6_9

compound backplate, easy loading/unloading, flexible and adjustable, etc. Generally, on the shop floor, fixtures are used to be designed by trial and error approaches, which in turn come out to be expensive and time-consuming [8]. Hence, the implementation of the finite element analysis (FEA) modeling approach in the design of fixtures would reduce unnecessary and uneconomical "trial and error" experimentation. Several researchers have designed and employed numerical fixture analysis for dynamics analysis [9–11]. This paper aims to design and develop a simplified, robust, and adjustable set-up fixture having features: precise and uniform clamping (top and side), proper accommodation of backplate, flexibility for thickness and length of workpieces, aligned tool travel along mating line, and structural stability at high downward pressure. SolidWorks software was being used to develop the 3D model. The model was simulated for four materials, i.e., Mild steel, Grey cast iron, Die Steel, and Hard alloy, in ANSYS software to analyze deformation behavior. The best material was chosen based on equivalent stress and deformation for the fabrication of the FSW fixture. The applicability of the designed fixture was analyzed on a vertical milling machine. Hardness measurements were done to assess the performance of the welded part.

2 Design of FSW Fixture

2.1 Prerequisite for the Design of a Fixture

In FSW, a non-consumable rotating tool with pin and shoulder is plunged between the workpiece plates and subsequently traversed along with the joint interface. Intense local heating of workpieces results from the friction between the tool and workpiece, which plasticizes the material around the tool. Further, the rotating tool traverses along the mating line and transfer the softened material around itself. Tool shoulder acts as a boundary to prevent the material from flowing out from the surface. Material's transportation took place from the front of the tool to the trailing edge, where it is forged into a joint [12–14]. Therefore for successful welding, TWI [6] suggested the following advisory for clamping and holding workpieces.

i. Rigid clamping of workpieces
ii. The adequate structural strength of all portions
iii. Back-plate for strong support to avoid bending
iv. Arresting of horizontal and vertical movements
v. Adequate heat sink for welding heat dissipation
vi. Stability during the process etc.

Failure of the abovementioned criterion would result in misalignment and buckling of the workpiece, improper mixing, plunging, tool path distraction, poor weld quality, etc.

2.2 Proposed Design

A simple and adjustable fixture had been designed to execute the FSW process on a vertical milling machine with considering all the necessities. Some innovation has been made to previously designed fixtures to adjust it on the milling machine's work bed.

Corners of the fixture were designed more filleted to withstand the high magnitude forces during processing. The extra cover had been added to the top faces' sides to avoid more deflection and to stabilize bending. The fixture's top and front views depicting the exact location of clamps and key slots are shown in Fig. 1a, b. Multiple and enough clamp length and number were considered to avoid buckling and facilitate uniform clamping. Thick fixture baseplate accommodation space had been provided to withstand high forces.

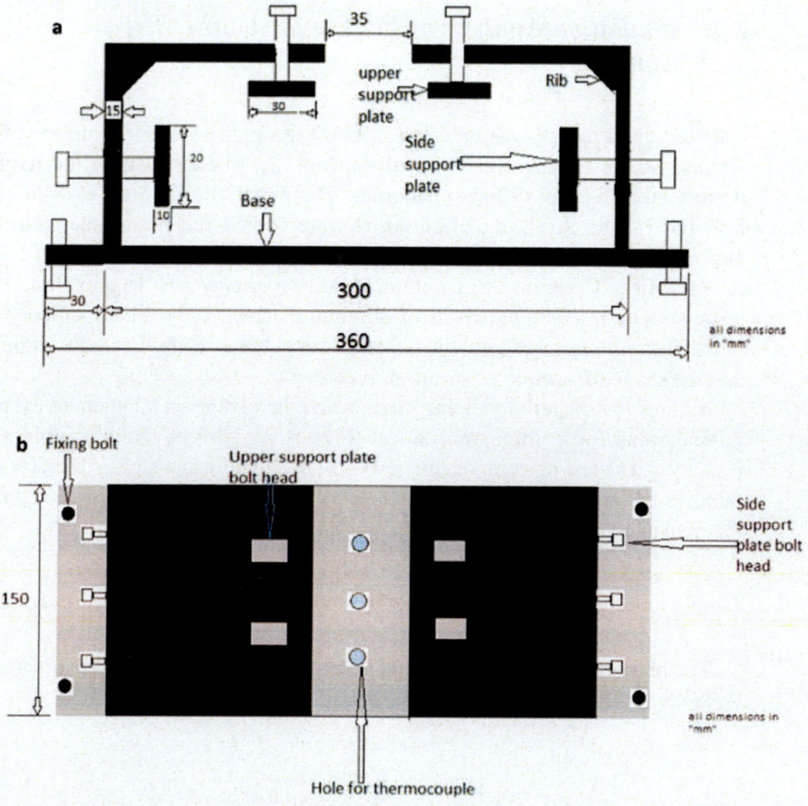

Fig. 1 Proposed FSW fixture design for vertical milling machine **a** Front view and **b** Top view

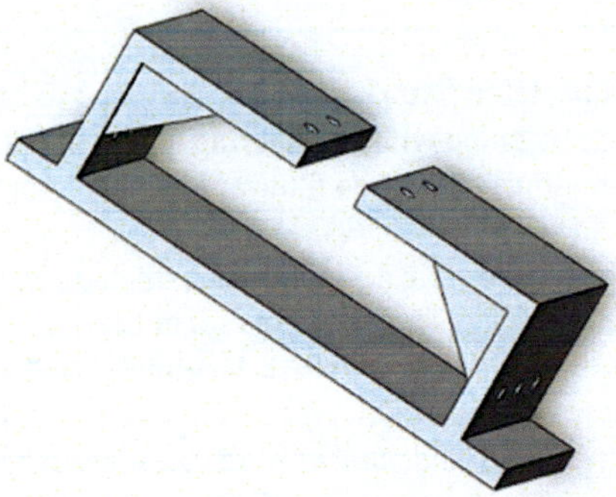

Fig. 2 SolidWorks model of the proposed design for analysis

2.3 Simulation Analysis of Fixture for Four Different Materials

This designed fixture was modeled in CAD designing software imported to finite element solver for deformation analysis (Fig. 2). The analysis of the fixture was performed using four different materials. The material Mild Steel (MS), Grey Cast Iron (GCI), Die Steel, and Hard Alloy were considered fixture materials for the analysis.

Von Mises Criterion of maximum distortion energy has been exploited for the static structural loading analysis of different materials. Von Mises's theory divides the total strain energy of a ductile material into volumetric (hydrostatic) strain energy and the shape (distortion or shear) strain energy.

It states that material yielding starts when the distortion component exceeds the critical value for a simple tensile test (Fig. 3). Von Mises stress combines all the Cauchy stress tensor components into a scalar quantity, which is directly comparable to the material's yield strength. The von Mises yield criterion expressed in mathematical form, as shown in Eq. 1.

$$\frac{1}{2}\left[(\sigma_1 - \sigma_2)^2 + (\sigma_2 - \sigma_3)^2 + (\sigma_3 - \sigma_1)^2\right] \leq \sigma_y^2 \quad (1)$$

The properties listed in Table 1 for considered four materials were obtained from ANSYS 15.0 standard material database.

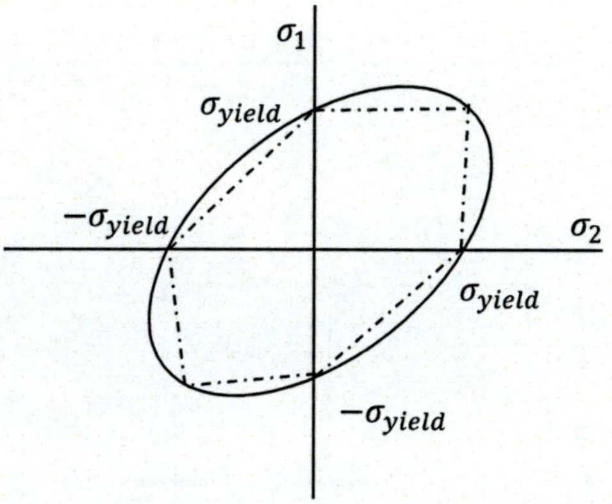

Fig. 3 Projection of the von Mises yield criterion

Table 1 Fixture material properties

Material properties	MS	GCI	DS	HA
Poisson ratio	0.29	0.29	0.28	0.28
Density (kg/m^3)	7850	7850	7700	7800
Young's modulus (GPa)	206	210	200	180
Tensile strength	550 N/mm^2	700 N/mm^2	670 N/mm^2	1110 m^2

3 Result of Deformation Analysis

Figures 4, 5, 6 and 7 shows the deformation and equivalent stress profile of Mild Steel, Grey cast iron, Die Steel, and Hard alloy, respectively.

From these contour plots, it can be observed that Mild Steel fixtures resulted in minimum deformation and comparable equivalent stress values (Table 2). In addition to it, Mild Steel is more commercially available than the other three ones. Hence based on both criteria, i.e., deformation analysis and commercial availability, Mild Steel finds out most suitable among the considered different materials.

4 Fabrication of MS Fixture

A single block of MS was taken for the fabrication of the fixture. With the application of different machining processes, the new design-based fixture was fabricated (Fig. 8) and used to execute FSW on a vertical milling machine.

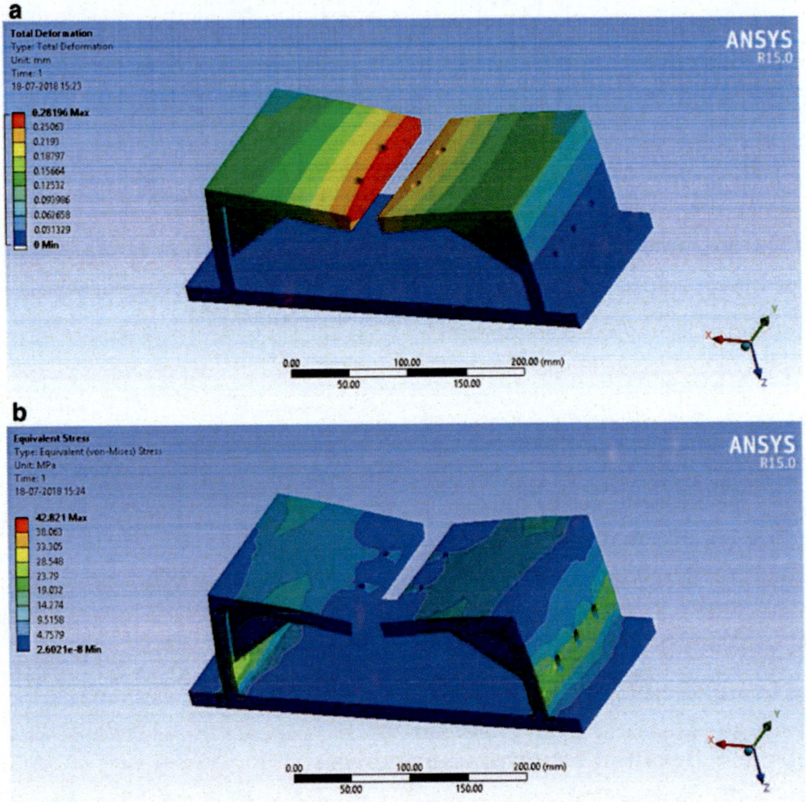

Fig. 4 Contour plot for GCI under loading condition **a** Deformation and **b** Equivalent stress

5 Practicability Test of the Developed Fixture

Commercially available pure rolled aluminum has been chosen for testing the fabricated fixture. The rolled plates were cut into rectangular pieces of 100 × 70 × 6 mm for performing FSW. Stainless steel tool (Grade-310, probe length 5 mm, $\varnothing_{\text{shoulder}}$ 18 mm) was used for the process. The workpieces were observed to be perfectly placed without any movement and gap between adjacent faces. No wrinkles were observed during clamping because of constant pressure at the joining line throughout and on both edges. Side and top supporting plate also increased the clamping area, which in turn exerted uniform pressure. The fixture base was kept sufficiently thick (35 mm) to avoid any distortion during welding runs. The fixture can be used for a wide range of dimensions of workpieces. Once the welding was over, the welded

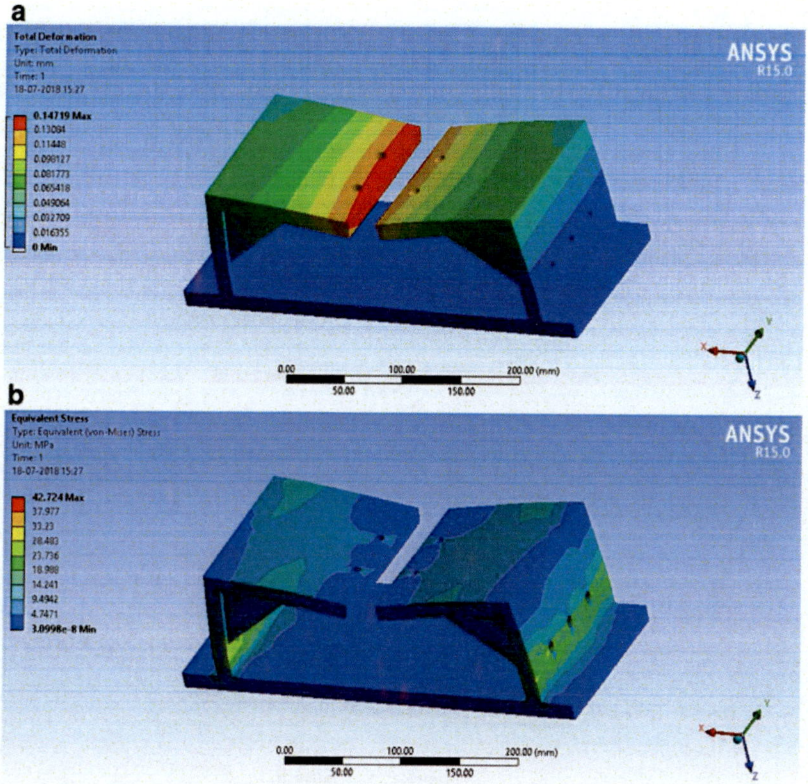

Fig. 5 Contour plot for MS under loading condition **a** Deformation and **b** Equivalent stress

specimens, as showed in Fig. 9b, were tested for hardness properties using a digital Rockwell hardness tester (Model—TRS—DM, Krutam Techno, India) as per the scheme mentioned in Fig. 9b.

The hardness plot (Fig. 10) shows that welded specimen has higher hardness values than the base metal hardness. Hence practical applicability of fabricated fixture has been proved.

6 Conclusions

In this paper, a new fixture design for FSW of Aluminum and its alloys was proposed, and simulated for four different structural materials, i.e., Grey cast iron, Mild Steel, Die Steel, and Hard alloy. The dynamics analysis showed better performance for Mild

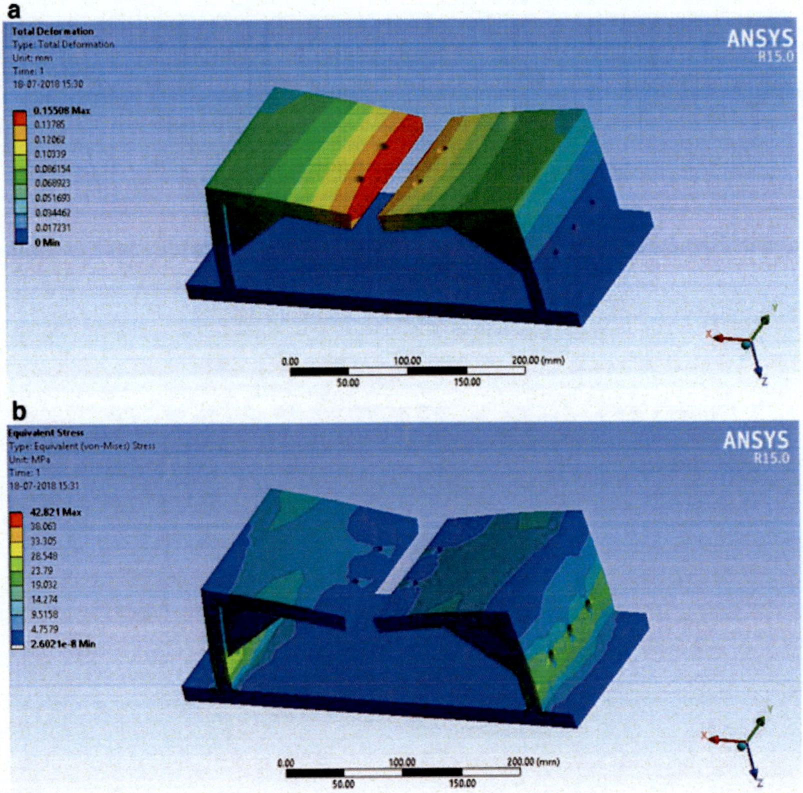

Fig. 6 Contour plot for DS under loading condition **a** Deformation and **b** Equivalent stress

Steel in terms of maximum deformation (0.14719 mm) and equivalent stress (42.724 MPa) as compared to GCI (0.28196 mm, 42.821 MPa), DS (0.15508 mm, 42.821 MPa), and HAS (0.17231 mm, 42.821 MPa). Based on the results and commercial availability, Mild Steel had been used to develop the fixture. Developed fixtures provide a flexible range with regard to dimensions and shape of workpieces as well tool profiles. Perfect clamping with zero gap and movement, uniformity in clamping pressure, and easy loading/unloading are the key features observed of fixture. Further, the experimental testing showed better FSWed joint and proved its practical applicability.

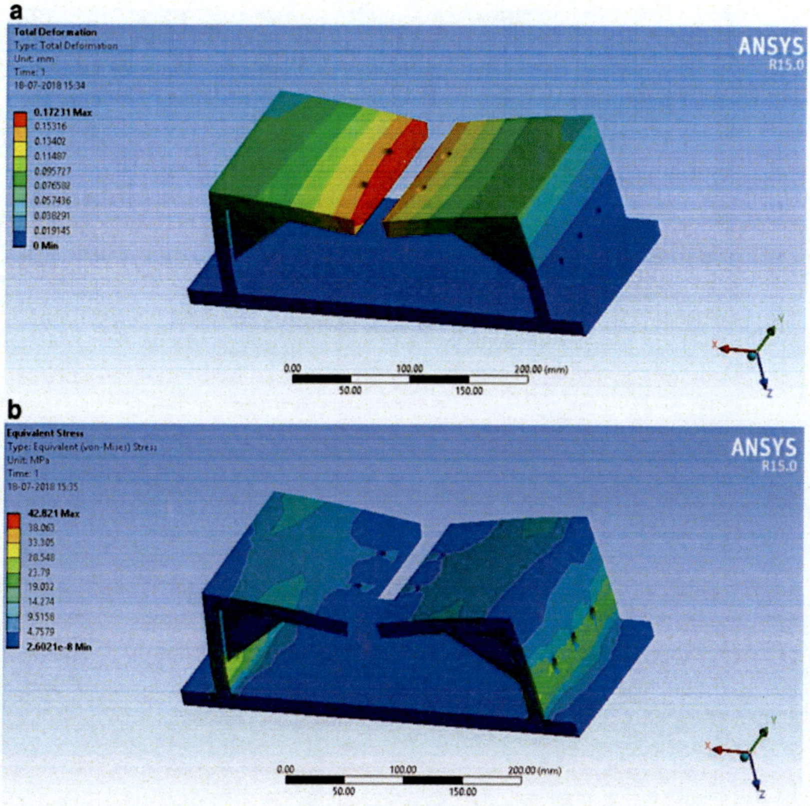

Fig. 7 Contour plot for HA under loading condition **a** Deformation and **b** Equivalent stress

Table 2 Results from the deformation analysis

Material	Total deformation (mm)	Equivalent (von Mises) stress (MPa)
Grey cast iron	0.28196	42.821
Mild steel	0.14719	42.724
Die steel	0.15508	42.821
Hard alloy	0.17231	42.821

Fig. 8 Fabricated fixture of Mild Steel for FSW on the vertical milling machine

b

Distance (mm)		Retreating side (RS)			Weld Center	Advancing side (AS)		
From the Top Surface	0 → 6	-15	-10	-5	0	5	10	15

Fig. 9 **a** Welded joint and **b** Hardness measurement scheme

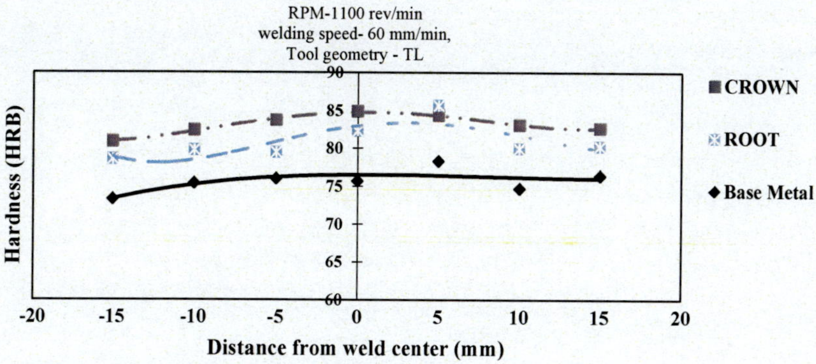

Fig. 10 Hardness plot for the welded joint specimen

References

1. Thomas WM, Nicholas ED, Needham JC, Murch MG, Templesmith P, Dawes CJ (1991) International Patent Application No PCT/GB92/02203 and GB Patent Application No 9125978.8
2. Magalhães VM, Leitão C, Rodrigues DM (2018) Friction stir welding industrialisation and research status. Sci Technol Weld Join 23(5):400–409
3. Thomas WM, Nicholas ED (1997) Friction stir welding for the transportation industries. Mater Des 18(4):269–273
4. Nandan R, DebRoy T, Bhadeshia HKDH (2008) Recent advances in friction-stir welding-Process, weldment structure and properties. Prog Mater Sci 53(6):980–1023
5. Kulekci MK, Esme U, Buldum B (2016) Critical analysis of friction stir-based manufacturing processes. Int J Adv Manuf Technol 85(5–8):1687–1712
6. Minton T, Mynors DJ (2006) Utilisation of engineering workshop equipment for friction stir welding. J. Mater. Process. Technol. 177(1–3):336–339
7. Titilayo AE, Makundwaneyi MD, Akinwale AS (2012) Reconfiguration of a milling machine to achieve friction stir welds. Appl Mech Mater 232:86–91
8. Mishra A, Rizvi S, Singh R (2017) Fabrication and vibration analysis on friction stir welding fixture for mass production. Mech Mech Eng 21(3):531–540
9. Fratini L, Micari F, Buffa G, Ruisi VF (2010) A new fixture for FSW processes of titanium alloys. CIRP Ann Manuf Technol 59(1):271–274
10. Arularasu S, Jothilingam A (2012) Design and development of low cost friction stir welding machine. In: IEEE-international conference on advances in engineering, science and management, ICAESM-2012, vol 3, no 12, pp 305–311
11. Ahmed S, Saha P (2018) Development and testing of fixtures for friction stir welding of thin aluminium sheets. J Mater Process Technol 252:242–248
12. Kah P, Rajan R, Martikainen J, Suoranta R (2015) Investigation of weld defects in friction-stir welding and fusion welding of aluminium alloys. Int J Mech Mater Eng 10(1)
13. Fuse K, Badheka V, Fuse K, Badheka V (2018) Bobbin tool friction stir welding : a review Bobbin tool friction stir welding: a review. Sci Technol Weld Join, 1–28
14. Eslami S, Tavares PJ, Moreira P (2016) Friction stir welding tooling for polymers: review and prospects. Int J Adv Manuf Technol

Experimental Investigation on Utilization of Waste Tire on Road Construction

Debjani Panda and Alisha Satapathy

1 Introduction

Increase in population demands overall development of new roads. As the vehicular traffic load is high in today's scenario, pavement failure is a major problematic situation. Consistent with that, large number of worn-out tires about 0.9 million tons or more are being generated in India which is a non-biodegradable material and that causes environmental pollution [1]. Therefore, it is necessary to reuse the waste tire with technical development in each field. In India, the road is a noteworthy vehicle framework and crucial economy which has the biggest and largest network. Flexible pavement designs are often utilized and comprises of major road network [2].

Some exhausted tires find their application in sports, recycled rubber tire to new tire, production of energy by burning for electric power, pulp and paper mill and cement in cement kiln, etc., and landfilling [3]. However, some parts of waste tires are discarded which cause various ecological issues. On the other hand, enlarged traffic variables like heavier axle burdens, surged traffic, and inflated tyres pressure require prevalent quality asphalt surfaces.

In order to mitigate the serious harm of pavement surfaces, if these worn-out tire materials can be utilized as an additive in the conventional mix design then the disposal problem will be reduced. In the present study, utilization of worn-out tires in road construction will be investigated and its suitability in terms of strength and ideal quantity is reported.

D. Panda (✉) · A. Satapathy
Parala Maharaja Engineering College, Berhampur, Odisha, India

R. Prasad et al. (eds.), *Advancement in Materials Processing Technology*,
Springer Proceedings in Materials 12,
https://doi.org/10.1007/978-981-16-3297-6_10

2 Literature Review

Scraped rubber is used in combination with bitumen (wet process) with changing size and quantity to produce modified bitumen. Coarser scraped rubber is used in the partial replacement of aggregates [4, 5].

Baraiya Niraj [4] in an experimental program on the waste rubber tire, mixed CR of 10–20% by weight of stone aggregate having size 22.4–6 mm with stone aggregate and bitumen at 160–170 °C without melting the CR in bitumen. An observation made was the reduction in the quantity of coarse aggregate by volume, increased strength and flexibility of highway and also gave better results in sub base with an optimum content of 5–20% of waste rubber.

Sharma et al. [6] modified VG30 by adding CR of 8, 10, 12, 14% by using wet process. It resulted in an increase in softening point, elastic recovery, and decrease in the penetration value. Also, Marshall Stability value increased by 18% up to 12% CR.

Rana [5] in his laboratory investigation on modified bitumen by using 15% CR by weight in different sizes of coarse, medium, fine, and superfine showed elevated softening point and reduced penetration value of CRMB with decreasing size of CR. Marshall test results showed 300–150 μ size CR in modified bitumen give more stability, minimum flow, and maximum density.

Patil et. al. [7] have illustrated the appropriate application of rubberized asphalt coating in hot blended blacktop and chipseal by leaving to react in heated state. There is a general advantage of reduction in noise, extended road maintenance, and accidents in rainy season due to improved traction between vehicle and asphalt.

Deshmukh Nitu and Kshirsagar [3] reported increase in softening point with addition of the increasing amount of waste rubber thereby making bitumen less susceptible to temperature changes subsequently giving protection against hot climatic conditions. Also from the ductility test, modified bitumen is harder in nature which would be suitable to gain stiffer bitumen asphalt. There is better adhesion between aggregate and binder due to which there is overall improved performance.

Sri and Daniel [8] in their research reported the effect of adding CR (1 and 2% of size #40 and #80) to asphalt mixture using wet process and the alluring test results of CRMB was within the standard necessity with a simultaneous improvement in strength and reduction in asphalt content and deformations.

3 Experimental Programme

3.1 Material

In the present experimental program, materials used are aggregate, bitumen, crumb rubber. Aggregates were collected from crusher, that contained coarse, fine, and filler stone dust. The collected stone dust was sieved in 20 to 4.75 mm, 2.36 to 75 μ,

Table 1 Properties of aggregates

Properties	Test method	Test value	Specification MORTH
Shape test	IS:2386-1963 (Part-1)	13.008% (F.I.) 23.780% (E.I.)	Not more than 35% combined
Impact test	IS:2386-1963 (Part-4)	20.089%	Max 27%
Abrasion test (Los-Angeles Abrasion)	IS:2386-1963 (Part-5)	25.6%	Max 35%
Specific gravity and water absorption	IS:2386-1963 (Part-3)	2.7 (sp.gravity) 1.2%(water absorption)	2.6–2.9 (sp.gravity) Max 2%(water absorption)
Crushing test	IS:2386-1963 (Part-4)	19.3%	Not more than 45%

Table 2 Properties of Bitumen

Properties	Test method	Test value	Specification IS:73-2013
Penetration test (25 °C, 100 g, 5 s, 0.1 mm)	IS:1203-1978	64	Not less than 45
Ring and Ball test, °C	IS:1205-1978	67.5	Not less than 47
Ductility test 27 °C, cm	IS:1208-1978	84.7	Not less than 75
Specific gravity test	IS:1202-1978	0.985	0.97–1.0

and pan categorized as coarse, fine, filler stone dust, respectively, and the desirable properties of aggregates are shown Table 1. Bitumen collected from Berhampur pwd office which properties are shown in Table 2. Crumb rubber was collected from waste tires recycling factory, Totapada, Khurdha, that contain all types of vehicle's tires (car, truck, bus, bike) and specific gravity is 2.07.

4 Methodology

4.1 Preparation of Modified Bituminous Concrete Mix

Collected aggregates were cleaned, oven dried, and graded as stated by MORT&H table 500–17. Factory processed crumb rubber was sieved to separate 300 μ size. The mentioned size CR of 10% by weight of aggregate was mixed with aggregate at a constant heated temperature of 160 °C.

Bitumen, heated to a pouring consistency temperature of around 140 °C was poured into the heated aggregate and crumb rubber mixture, 5% weight of aggregate so as to bind the mixture. A uniform bitumen coating on aggregate and crumb rubber was achieved by thorough mixing maintaining a temperature of 180 °C. The above technique was recurrent with 0.5% increment of bitumen by weight of aggregate upto 6.5%. Likewise, 10 and 5% CRMM was prepared using the same varying percentage of bitumen.

4.2 Preparation of Specimen

For preparation of specimen, the standard size cylindrical mold and rammer were cleaned and a little glycerine was applied to it. Then the mixture was transferred to the mold and tampered by using a trowel and levelled the surface. Keeping in view higher traffic flow conditions, the specimen was densely compressed by applying 75 blows on each side manually using a rammer after which the specimen was drawn out and rested for at least 6 h.

4.3 Testing the Mix Design of Specimen

After preparation and extraction of specimens, before testing their thickness, weight in air, weight in water was measured. Marshall Apparatus was used to determine stability and flow value at 60 °C as per standard procedure. Temperature conditioning of the specimen was done at 60 °C water bath for 30–45 min followed by testing, within 3–5 min after taking out from the water bath.

5 Result and Discussion

Marshall Quotient is the relation between stability and flow value, the ratio of which, if high, indicates a stiffer mix, i.e., more resistance to deformation which can be permanent in nature [9] (see Table 3).

From Fig. 1, it is seen that stability or strength of mixes exhibit increasing trend up to 5.5% binder content then decreases. The voids of the dry mix of aggregates and crumb rubber are filled with bitumen which is a binder. With the gradual increment of binder, more fluidity of the mix causes decrease in strength. Also as compared to conventional mix, the stability of 10% CRMM is low comparatively but reduction in quantity of CR from 10 to 5% demonstrates more stability value, even more than conventional mix.

From Fig. 2 we observed higher deformation in 10% CRMM in contrast to 5% CRMM and conventional mix as the flowability property increases dramatically with increase in the quantity of bitumen in bituminous mixes.

In Fig. 3, it is seen that the air voids are much higher in 10% CRMM. The probable cause of such nature is improper interlocking among the materials although air voids

Table 3 Marshall mix test result

Mix type	Binder content (%)	Stability (kN)	Flow value (mm)	Air voids in mix (%)	VMA (%)	VFB (%)	Specific gravity (g/cc)	Marshall quotient (kN/mm)
Conventional mix	5	13.754	3	7.235	18.068	59.953	2.27	4.584
	5.5	24.181	3.3	5.183	17.263	69.972	2.30	7.327
	6	10.530	4	3.74	17.009	78.008	2.32	2.632
	6.5	9.709	5	3.962	18.244	78.280	2.30	1.941
5% CR modified mix	5	14.866	2.7	5.650	18.020	64.012	2.272	5.505
	5.5	29.314	3.1	4.744	17.706	72.269	2.291	9.456
	6	11.508	3.8	4.105	17.862	78.351	2.297	3.028
	6.5	9.883	4.6	3.833	18.940	79.737	2.288	2.148
10% CR modified mix	5	5.480	5.16	9.190	18.921	51.426	2.188	1.062
	5.5	8.769	5.2	8.269	18.921	56.298	2.1952	1.686
	6	4.383	6.2	7.482	19.103	60.830	2.201	0.707
	6.5	4.383	6.4	6.966	19.456	64.193	2.195	0.684

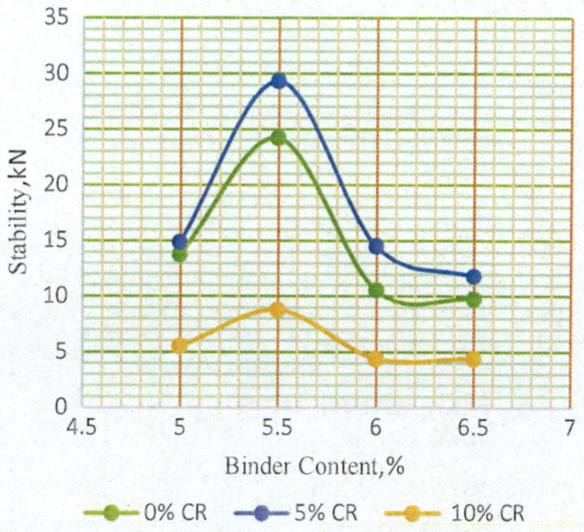

Fig. 1 Stability versus binder content

decrease with increase in binder contents which occupy the void spaces. Alternatively, it can be said that 10% CRMM is not an ideal modification. In bituminous concrete mixes, ideally 4% air voids are considered as too many air voids cause disintegration of mixes. Also, limited air voids in the mixes cause bleeding of bitumen when allowed for traffic movement of various loads. Bitumen content corresponding to 4% air voids for conventional mix and 5% CRMM is 5.85% and 6.13%, respectively.

It is reconnoiter from Fig. 4, the specific gravity of conventional mix increases up to a certain point after which it decreases. Such characteristic is possible due to change in binder content and air voids. It is observed from the graph that increase in

Fig. 2 Low value versus binder content

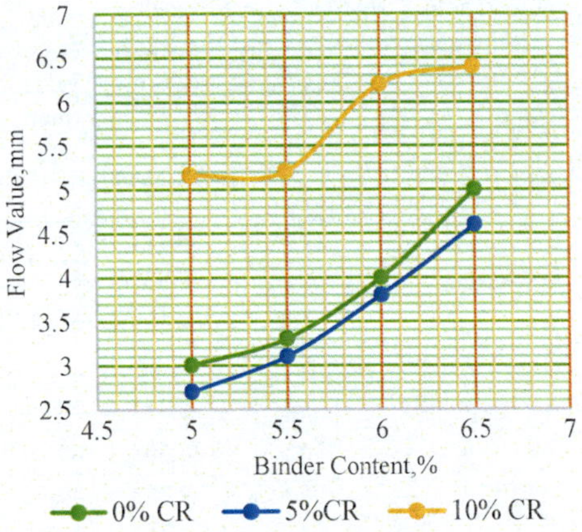

Fig. 3 Specific gravity versus binder content

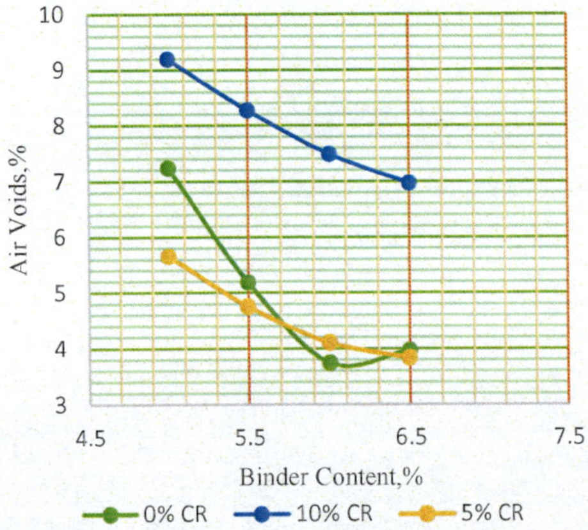

Fig. 4 Air voids versus binder content

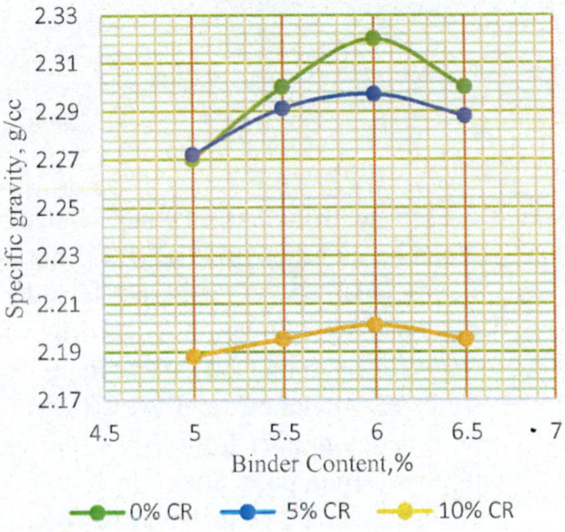

the amount of CR in the mix decreases specific gravity while the nature of the graph is the same. The reason behind is the addition of CR which has low specific gravity which overall reduces the specific gravity of the mix.

Computation of optimum binder content (OBC) is average of bitumen binder content analogous to highest stability, density, and 4% air voids. From present study, we observed that OBC of conventional mix and 5% CRMM is 5.826% and 5.81%, respectively, with stability values 16.2 kN and 21.1 kN.

6 Conclusion

The stability value of 5% CRMM was increased by 8.08–21% as compared to conventional mix. In comparison between conventional mix and 5% CRMM, the flow value was decreased from 10 to 8% and voids space was more. 5% CRMM has high computed Marshall Quotient. Although, there is not much difference in the OBC of conventional mix and 5% CRMM but the stability at OBC increased by 21.2% and specific gravity decreased by 1.29%. 10% CRMM stability reduced to 63.73%.

It can be inferred that 10% CR in bituminous roads has no potential to greatly improve the marshall performance in terms of strength and deformation with least Marshall Quotient. OBC in 5% CRMM reduced by 0.17% only although it can be cost-effective for voluminous road construction being a partial substitute of bitumen which is expensive. Thus, addition of 5% crumb rubber of sieved size 300 μ can be put to use and can boost the road efficiency in aspects of strength and deformations. Utilization of waste tire is lessening environmental problem to some level.

References

1. Madona L, Mejdi J (2016) Thermochemical conversion of waste tyres—a review. Environ Sci Pollut Res. https://doi.org/10.1007/s11356-016-7780-0
2. Sudharshan RB, Venkata Hussain RN (2016) Performance evaluation of crumb rubber modified bitumen by using various sizes of crumb rubber. Int J Res Appl Sci Eng Technol (IJRASET) 4(III):2321–9653
3. Deshmukh Nitu H, Kshirsagar DY (2017) Utilization of rubber waste in construction of flexible pavement. Int J Adv Res Dev 2(7)
4. Baraiya Niraj D (2013) Use of waste rubber tyres in construction of bituminous road. Int J Appl Innovat Eng Manag (IJAIEM) 2(7). ISSN 2319-4847
5. Rana MN (2014) A study on the performance of crumb rubber modified bitumen by varying the sizes of crumb rubber. Int J Eng Trends Technol (IJETT) 14(2)
6. Kumar SP, Kumar SA, Arora TR (2013) Experimental study of flexible pavement by using waste rubber tyre. Int J Eng Res Technol (IJERT) 2(8). ISSN 2278-0181
7. Patil SB, Lole AA, Bavane NU, Shinde SS (2016) Use of waste tyres in road construction. Int Res J Eng Technol (IRJET) 3. e-ISSN 2395-0056. p-ISSN 2395-0072
8. Sri WP, Daniel T (2017) Uses of crumb rubber as an additive in asphalt concrete mixture. Sci Direct Sustain Civ Eng Struct Constr Mater 171:1384–1389. https://doi.org/10.1016/j.proeng.2017.01.451

9. Lakshman SK, Debjani P (2019) Study on strength characteristics improvement of polyethylene modified bituminous concrete mixes. Key Eng Mater 803: 216–221. ISSN 1662-9795. https://doi.org/10.4028/www.scientific.net/KEM.803.216. 2019 ©Trans Tech Publications Ltd., Switzerland.
10. Ministry of Road Transport and Highway, Specification for road and bridge works. Fifth revision

Study of Mechanical Properties of Chemically Treated Kenaf Fiber and Its Composites

Sudhakar Behera, Rakesh Kumar Gautam, and Sunil Mohan

1 Introduction

The use of composites reinforced with natural fibers is gaining momentum because of their biodegradability, ease of availability, lightweight, etc. [1]. Cellulose, hemicellulose, and lignin are the major chemical constituents of natural fiber. But these natural fibers have poor compatibility with the polymer matrix as the natural fibers are water absorbent in nature. Researchers have done various chemical treatments on the natural fibers to reduce their water retention ability and to make the fiber surface rough for better compatibility of the fiber and the matrix [2–4].

Kenaf fiber is easily available in India and can be used as a reinforcing material in making composites because of their high strength, low density, non-toxicity, etc. Manral et al. [5] studied the effect of sodium acetate treatment on the non-cellulosic content of kenaf fiber. The study concluded that the above chemical treatment has reduced the non-cellulosic content of the fiber and increased its crystallinity index. Zhu et al. [6] found that the content of cellulose increases while the hemicellulose and lignin content of sisal fiber decreases with an increase in fiber soaking time. They also concluded that with an increase in treatment time, the tensile strength also increases although the tensile strength of untreated fiber was higher compared to untreated fiber. Dashtizadeh et al. [7] investigated the influence of NaOH and silane modification on the tensile and interfacial shear strength of kenaf fiber. They found that 2% NaOH treatment of kenaf for 4 h has considerably increased the tensile and interfacial strength of the kenaf fiber. Hasan et al. [8] found that the thermal stability, tensile modulus, and stiffness of the kenaf fiber have increased after alkali treatment but there was a slight decrease in the tensile strength value of the fiber. Wu et al. [9]

S. Behera (✉) · R. K. Gautam · S. Mohan
Department of Mechanical Engineering, Indian Institute of Technology (BHU), Varanasi 221005, Uttar Pradesh, India
e-mail: sudhakarbehera.rs.mec18@itbhu.ac.in

R. Prasad et al. (eds.), *Advancement in Materials Processing Technology*, Springer Proceedings in Materials 12,
https://doi.org/10.1007/978-981-16-3297-6_11

studied the effect of NaOH, silane, tetraethylorthosilicate (TEOS) sol–gel, TEOS, and silane combination treatment on the sisal fiber. The study concluded that the heat resistance and tensile strength have been enhanced by the chemical modification of the sisal fiber and the best result was shown by combination treatment. Ibrahim et al. [10] analyzed the impact of various concentrations of NaOH modification on the tensile behavior of a single kenaf fiber. The study found that the 6% NaOH treatment of kenaf fiber showed the best mechanical properties.

In the last decade, many experiments were performed to chemically modify the kenaf fiber but few experiments were performed that involve $NaHCO_3$ and benzoyl chloride to treat the kenaf fiber. So, in this experiment, the kenaf fiber is subjected to $NaHCO_3$ and benzoyl chloride treatment to improve its physical and mechanical properties. Scanning electron microscope (SEM) was used to study the surface of the fiber. Microdroplet tests and tensile strength tests were done to study their interfacial and tensile strength. Compositional analysis of the fiber was also done to study its chemical constituents.

2 Materials and Methods

2.1 Materials

The polylactic acid (PLA) polymer used in this experiment was supplied by Sigma-Aldrich. The processing temperature of PLA is in the range of 160–240 °C. Kenaf fiber bundles were supplied by Mohanty Textiles, Cuttack, Odisha, India. $NaHCO_3$, NaOH, and benzoyl chloride were provided by Ajay Chemicals, Varanasi, India.

2.2 Fiber Surface Treatments

Initially, the kenaf fiber bundles were cut into 45 mm length and were cleaned with distilled water to remove the contaminants. These untreated (UT) fibers were then dried in the sunlight for 6 h.

2.3 $NaHCO_3$ Treatment

The UT kenaf fibers were dipped in a solution of 11 wt % $NaHCO_3$ (room temperature) for a time period of 24 h, then cleaned with deionized water till the pH become 7. $NaHCO_3$ treated (SBT) fibers were dried in an oven for 28 h at 45 °C.

2.4 *Benzoyl Chloride Treatment*

The UT kenaf fibers were first suspended in a solution of 18 wt % $NaHCO_3$ for 4 h and then dipped in a solution of benzoyl chloride for 20 min. Benzoyl chloride treated (BCT) fibers were cleaned with distilled water, and then with ethanol, and placed in an oven for drying at 75 °C for 5 h.

2.5 *Composition Analysis*

Neutral detergent solution and acid detergent solution techniques were employed to find out the cellulose, hemicellulose, and lignin content of kenaf fiber [7]. This experiment was done at the Institute of agricultural science, Banaras Hindu University, India.

2.6 *Scanning Electron Microscopy (SEM) Analysis*

The surface morphology of treated and untreated kenaf fiber was investigated with the help of SEM. SEM analysis was done in a JSM-7900F field emission SEM (JEOL Ltd.).

2.7 *Tensile Strength of Kenaf Fiber*

An Instron universal testing machine (model 5900) was used to determine the tensile strength of individual kenaf fiber at a loading speed of 0.5 mm/min. The ASTM standard chosen was ASTM C 1557–03. A digital microscope was used to obtain the fiber diameter. The tensile strength (TS) of the fiber can be obtained from the following formula (Eq. 1).

$$\mathrm{TS} = \frac{F_t}{\pi r^2} \tag{1}$$

Here, F_t is designated as the maximum tensile force and r is designated as the radius of the kenaf fiber.

2.8 Microdroplet Debonding Test

Melting of PLA was done in an oven for 10 min at a temperature of 100 °C. A single kenaf fiber was attached to the melted PLA filaments and was put in an oven at 190 °C to prepare the samples for the test. An Instron machine (model 5900) was used to perform the debonding test at a pull-out velocity of 0.5 mm/min. Interfacial strength (IS) can be obtained from the following formula (Eq. 2).

$$\mathrm{IS} = \frac{F_{pl}}{LC} \tag{2}$$

Here, F_{pl} is designated as maximum peak load, L is known as the depth of the fiber in the microdroplet, and C is called the cross-sectional perimeter of kenaf fiber.

3 Results and Discussion

3.1 Surface Morphology of Kenaf Fiber

SEM micrographs of untreated, SBT, and BCT kenaf fibers were shown in Fig. 1. The presence of amorphous waxy impurities and the packed structure of the fiber were clearly visible in the SEM image of UT fiber (Fig. 1a). This type of impurities results in improper interfacial bonding. However, both the $NaHCO_3$ and benzoyl chloride treatment has considerably removed the impurities from the fiber surface (Fig. 1a, b). Fiber surface modification has also led to fiber fibrillation and reduction in fiber diameter which may have resulted in good interfacial bonding between the matrix and the fiber. Comparable conclusions were also reported for other modified natural fibers [11, 12].

3.2 Effect of Chemical Treatment on the Composition of Kenaf Fiber

The compositional analysis of the kenaf fiber before and after the chemical treatment was given in Table 1. It can be observed from Table 1 that both the chemical treatment has significantly changed the chemical composition of the kenaf fiber. Both the $NaHCO_3$ and benzoyl chloride treatment was responsible for the loss of hemicellulose and lignin content of the fiber. However, cellulose content increases with $NaHCO_3$ and benzoyl chloride treatment. Reduction of hemicellulose content is more compared to lignin because hemicellulose consists of many hydroxyl and acetyl groups which can be easily removed by chemical treatment. But it is quite difficult to reduce the lignin content because the macromolecular structure of lignin

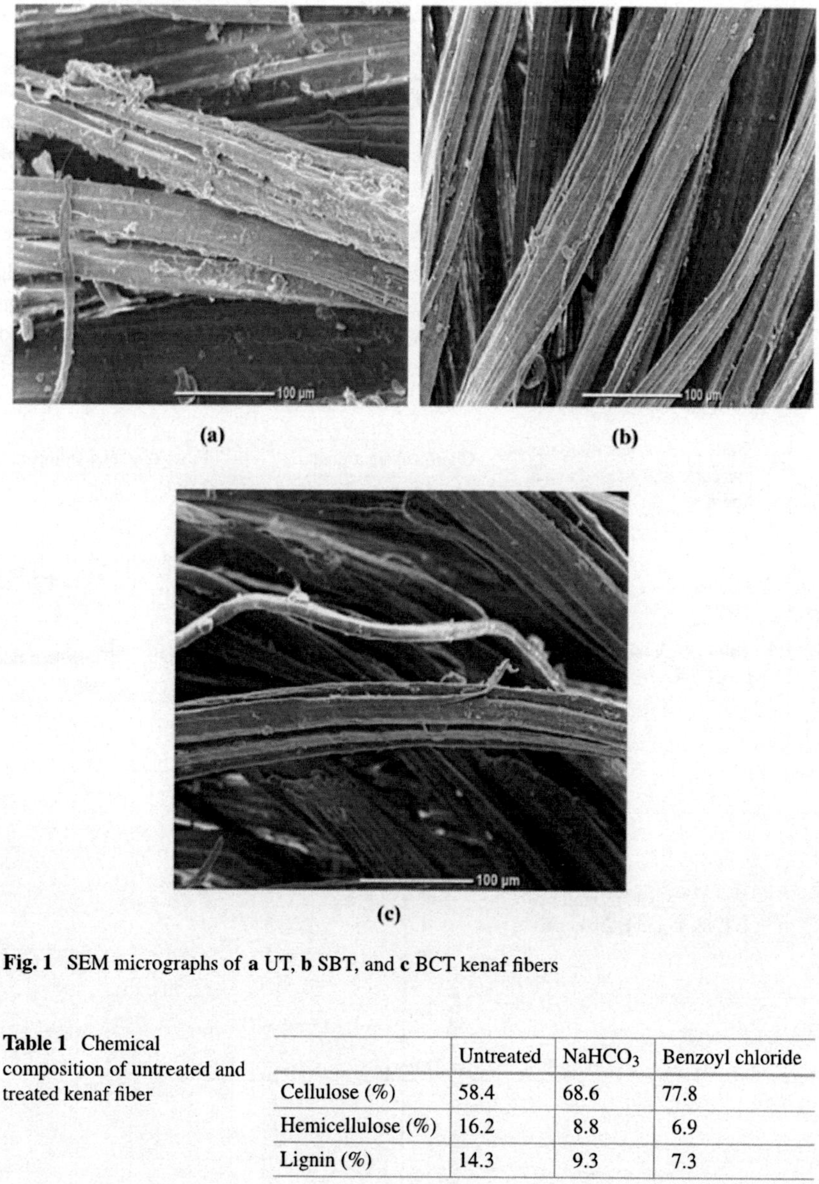

Fig. 1 SEM micrographs of **a** UT, **b** SBT, and **c** BCT kenaf fibers

Table 1 Chemical composition of untreated and treated kenaf fiber

	Untreated	$NaHCO_3$	Benzoyl chloride
Cellulose (%)	58.4	68.6	77.8
Hemicellulose (%)	16.2	8.8	6.9
Lignin (%)	14.3	9.3	7.3

is very strong as it provides structural support to the plant fibers. Benzoyl chloride treatment was responsible for lower hemicellulose and lignin content and higher cellulose content than $NaHCO_3$ treatment. Decrease in hemicellulose and lignin content is also responsible for lower water absorption behavior of natural fiber.

3.3 Tensile Properties of Untreated and Treated Kenaf Fibers

Tensile properties are one of the major factors that determine the strength and stiffness of the fiber. Prior to the tensile test, the diameter of the raw and chemically modified kenaf fibers were determined with help of a digital optical microscope and were represented in Table 2. It can be seen that the chemically modified fibers have smaller average diameter compared to the untreated fibers. The tensile strength and tensile modulus of untreated and modified kenaf fibers were given in Table 3. The highest tensile strength was obtained for UT kenaf fiber and the highest tensile modulus was obtained for BCT kenaf fibers (Table 3). From Fig. 2, it can be seen that the chemically modified fibers have comparatively higher tensile modulus than the untreated fiber. BCT kenaf fiber has the highest tensile modulus followed by SBT fiber and UT fiber.

Table 2 Average diameter of untreated and treated kenaf fibers

Chemical treatment	Fiber diameter (micrometer)
Untreated	90.43 ± 20.33
$NaHCO_3$ treatment	79.86 ± 12.23
Benzoyl chloride treatment	70.23 ± 15.16

Table 3 Tensile properties of kenaf fiber

Chemical treatment	Tensile strength (Mpa)	Tensile modulus (Gpa)
Untreated	178.25	23.67
$NaHCO_3$ treatment	157.62	26.13
Benzoyl chloride treatment	131.12	30.25

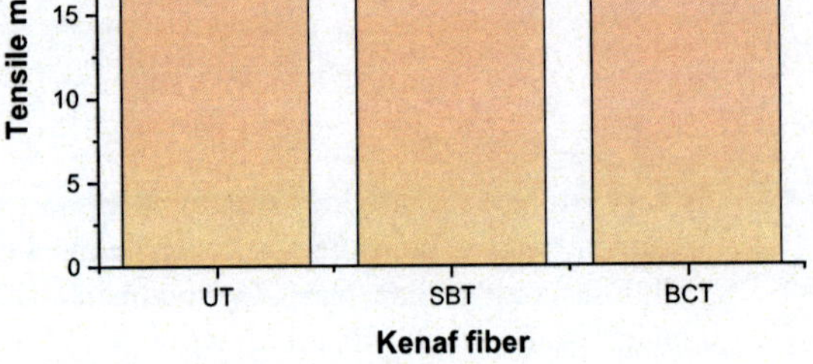

Fig. 2 Tensile modulus of UT, SBT, and BCT kenaf fiber

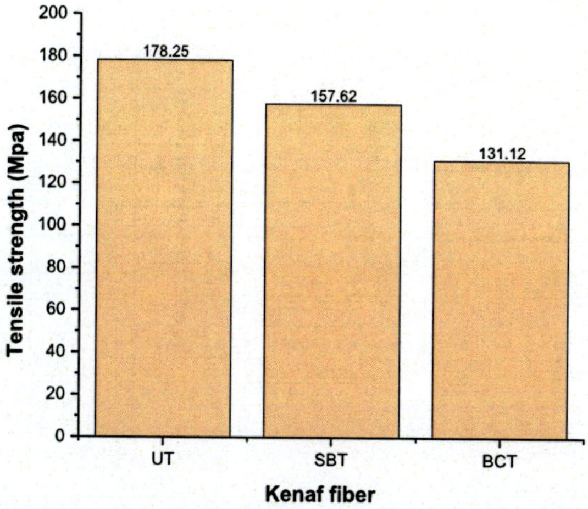

Fig. 3 Tensile strength of UT, SBT, and BCT kenaf fiber

This is due to the reduction in lignin content by the chemical treatment which makes the fiber more crystalline and stiffer than the untreated fibers [13]. However, the treated kenaf fiber has comparatively lower tensile strength than the untreated fiber (Fig. 3). UT kenaf fiber has the highest tensile strength followed by SBT fiber and BCT fiber. This reduction in the tensile strength of chemically modified fiber may be due to the structural damage of the fiber caused by the chemical treatment [14].

3.4 Interfacial Strength of Kenaf Fiber and PLA

Interfacial strength is one of the major factors that govern the mechanical behavior of fiber-reinforced composites. The interfacial strength of untreated and treated kenaf fiber with PLA was represented in Fig. 4. It can be observed that both $NaHCO_3$ and benzoyl chloride treated kenaf fiber has an improved interfacial strength compared to the untreated kenaf fiber. The best interfacial strength was shown by BCT fiber while the lowest interfacial strength was shown by UT kenaf fiber and the intermediate interfacial strength was shown by SBT fiber. This happens because the chemical treatment was responsible for the reduction in hemicellulose and lignin content and BCT fiber has the highest reduction in hemicellulose and lignin content. The reduction in hemicellulose and lignin content has also reduced the water absorption capacity of the fiber and hence increased the interfacial strength of treated kenaf fiber and PLA [15, 16].

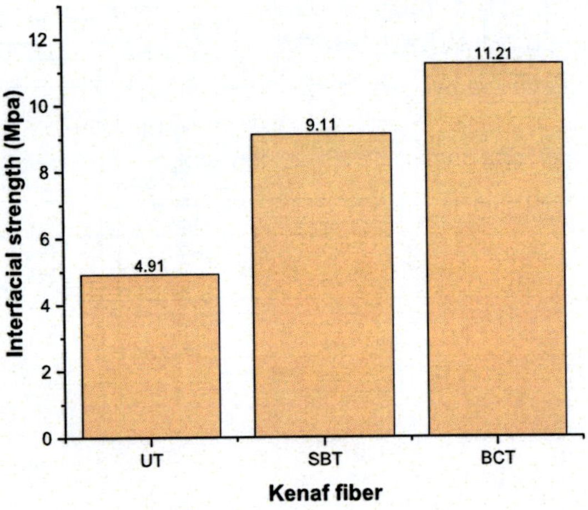

Fig. 4 Interfacial strength of UT, SBT, and BCT kenaf fiber and PLA matrix

4 Conclusions

Chemical composition, surface morphology, tensile properties, and interfacial strength of UT, SBT, and BCT kenaf fiber and its PLA composites were studied in this research. SEM analysis of both the chemically treated fiber surface confirmed the partial removal of biological dirt from the kenaf fiber surface.

Both $NaHCO_3$ and benzoyl chloride treatment have increased the cellulose content and decreased the content of hemicellulose and lignin in the fiber. BCT kenaf fiber has the highest cellulose content and lowest hemicellulose and lignin content.

Chemical modification has reduced the tensile strength of kenaf fiber. UT kenaf has the highest and BCT fiber has the lowest tensile strength. But chemical modification has increased the tensile modulus of kenaf fiber. BCT fiber has the highest and UT fiber has the lowest tensile modulus.

Chemical treatment has triggered the improvement of the interfacial strength of kenaf fiber and PLA biodegradable matrix. The best interfacial strength was shown by BCT kenaf fiber.

References

1. Ramamoorthy SK, Bakare F, Herrmann R, Herrmann R, Scrifvas M (2015) Performance of biocomposites from surface modified regenerated cellulose fibers and lactic acid thermoset bioresin. Cellulose 22:2507–2528. https://doi.org/10.1007/s10570-015-0643-x
2. Nopparut A, Amornsakchai T (2016) Influence of pineapple leaf fiber and its surface treatment on molecular orientation in, and mechanical properties of, injection molded nylon composites. Polym Testing 52:141–149. https://doi.org/10.1016/j.polymertesting.2016.04.012

3. Rong MZ, Zhang MQ, Liu Y, Yang GC, Zheng HM (2001) The effect of fiber treatment on the mechanical properties of unidirectional sisal-reinforced epoxy composites. Compos Sci Technol 61:1437–1447. https://doi.org/10.1016/S0266-3538(01)00046-X
4. Soares FA, Nachtigall SMB (2013) Effect of chemical and physical foaming additives on the properties of PP/wood flour composites. Polym Testing 32(4):640–646. https://doi.org/10.1016/j.polymertesting.2013.02.009
5. Manral A, Bajpai PK (2020) Analysis of properties on chemical treatment of kenaf fibers. Materials today: Proceedings. https://doi.org/10.1016/j.matpr.2020.03.266
6. Zhihua Z, Hao M, Zhang N (2020) Influence of contents of chemical compositions on the mechanical property of sisal fibers and sisal fibers reinforced PLA composites. J Nat Fibers 17(1):101–112. https://doi.org/10.1080/15440478
7. Dashtizadeh Z, Abdan K, Jawaid M, Khan MA, Behmanesh M, Dashtizadeh M, Francisco C, Ishak M (2016) Effect of chemical treatment on kenaf single fiber and bio-phenolic resin regarding its tensile and interfacial shear stress. Middle-East J Sci Res 24(9):2685–2692. https://doi.org/10.5829/idosi.mejsr.2016.24.09.23930
8. Hassan A, Isa MRM, Ishak ZAM, Ishak NA, Rahman NA, Salleh FM (2018) Characterization of sodium hydroxide-treated kenaf fibres for biodegradable composite application. High Perform Polym 30(8):890–899. https://doi.org/10.1177/0954008318784997
9. Wu M, Sun Z, Zhao X (2020) Effects of different modification methods on the properties of sisal fibers. J Nat Fibers 17(7):1048–1057. 1080/15440478.2018.1554517
10. Ibrahim MI, Dolah R, Yusoff MZM, Salit MS, Hassan MZ (2017) Chemical treatment evaluation of tensile properties for single Kenaf fiber. J Adv Res Appl Mech 32(1):9–14
11. Sreekala MS, Thomas S (2003) Effect of fibre surface modification on water-sorption characteristics of oil palm fibres. Compos Sci Technol 63(6):861–869. https://doi.org/10.1016/S0266-3538(02)00270-1
12. Bledzki AK, Gasssan J (1999) Composites reinforced with cellulose based fibres. Prog Polym Sci 24:221–274. https://doi.org/10.1016/S0079-6700(98)00018-5
13. Kargarzadeh H, Ahmad I, Abdullah I, Dufresne A, Zainudin SY, Sheltami RM (2012) Effects of hydrolysis conditions on the morphology, crystallinity, and thermal stability of cellulose nanocrystals extracted from kenaf bast fibers. Cellulose 19:855–866. https://doi.org/10.1007/s10570-012-9684-6
14. Zhong J, Zhang L, Yu J, Tan TW, Zhang X (2010) Studies of different kinds of fiber pretreating on the properties of PLA/sweet sorghum fiber composites. J Appl Polym Sci 117:1385–1393. https://doi.org/10.1002/app.31925
15. Rozman HD, Tan KW, Kumar RN, Abubaker A, Ishak ZAM, Ismail H (2000) The effect of lignin as a compatibilizer on the physical properties of coconut fiber Polypropylene composites. Eur Polymer J 36:1483–1494. https://doi.org/10.1016/S0014-3057(99)00200-1
16. Thielemans W, Wool RP (2004) Butyrated kraft lignin as compatibilizing agent for natural fiber reinforced thermoset composites. Compos Part A Appl Sci Manuf 35:327–338. https://doi.org/10.1016/j.compositesa.2003.09.011

Study of Structure and Microstructure Evolution in Iron-Rich Aluminum Alloy Cast Through Non-equilibrium Processes

Pritiman Mohapatra, Rajat Roy, Manish Kumar Soni, and B. Ravi Kumar

1 Introduction

In recent years, the transport industries have shown incredible interest to increase their efficiency of vehicles especially mileage by reducing their body weight. Therefore, it has drawn great attention to the scientists and researchers to the development of high-strength lightweight alloys that can be used to replace the structural components of the vehicles. Aluminum is one of the most abundant and low-cost elements available on Earth. Aluminum and its alloys have very wide applications in various industries because of its lightweight property. The range of tensile strength achieved in wrought aluminum alloys is up to 600 MPa through the precipitation hardening processes. On the other hand, this strength is considerably very less at an elevated temperature of 423 K and for this reason, its applications in structural components are very limited. At elevated temperatures, the strength of the alloy can be enhanced by developing a new processing route or through alteration of structure and microstructure by heat treatment or through phase transformation. In industrial aluminum alloys, iron and silicon are found as the most common impurities, at the same time, iron in the aluminum matrix increases the strength of the alloy. By this virtue, we can add maximum iron to the aluminum matrix to achieve the required strength. Since the solubility of iron in the aluminum matrix under equilibrium conditions is very low, i.e., <0.03 wt.% [1–3], it forms hard and brittle intermetallics phase $Al_{13}Fe_4$ at all ranges of iron contents. The formability of the materials decreases drastically due to the presence of the $Al_{13}Fe_4$ phase. Therefore, non-equilibrium processing routes such as rapid solidification processing (RSP) [5–7], mechanical alloying (MA) [5, 6, 8–10], electron beam deposition (EBD), and severe plastic deformation (SPD) [5–7,

P. Mohapatra
Malaviya National Institute of Technology, Jaipur 302017, India

R. Roy · M. K. Soni (✉) · B. Ravi Kumar
CSIR, National Metallurgical Laboratory, Jamshedpur 831007, India

R. Prasad et al. (eds.), *Advancement in Materials Processing Technology*,
Springer Proceedings in Materials 12,
https://doi.org/10.1007/978-981-16-3297-6_12

11–13] can be chosen for extending the solubility of iron in the aluminum matrix. In addition to this, the non-equilibrium processing route also enhances the strength of the system by refining microstructures down to the nanometer regime and lead to the formation of stable and metastable phases. Processing routes such as high-pressure torsion (HTP) and equal channel angular pressing (ECAP) increased the strength of the system at the same time the subsequent aging further increases the strength of the system [12, 13]. The combination of high-pressure torsion and rapid quenching showed an increased concentration of iron content in the aluminum matrix [6, 7]. Further, refinement of the alpha phase aluminum grains and second phase particles (Al13Fe4) in the matrix showed an increase in mechanical properties of the system such as tensile strength and ductility [13].

The present investigation is an effort to understand the effect of non-equilibrium processes during casting of Al-10Fe wt.% alloy and the evolution of its structure and microstructure. This particular study deals with the suction casting of Al–Fe alloy followed by annealing at various parameters and conditions of heating and cooling rates. The main objective was to study microstructure alterations of the alloy at high Iron content under non-equilibrium processing conditions.

2 Material and Methods

2.1 *Alloy Preparation*

Aluminum-based Al–Fe alloy was prepared with an Iron content of 10 wt.% by electric arc melting and remelting several times to achieve chemical homogeneity in an inert atmosphere of Argon. This as-cast ingot was then subsequently suction cast into 2 mm strips. The suction cast strips were further subjected to an annealing treatment at ultra-fast heating and cooling rates in a custom-designed annealing simulator. The heating rate was kept at 10°/s and the cooling rate at 100°/s as well as the isothermal annealing time of all annealing experiments was 60 s. Then and there the temperatures chosen for experiments were 750, 800, and 825 °C which was well above the melting point of Aluminum.

2.2 *Microstructure and Structure*

As-cast, suction cast, and annealed samples were prepared through standard metallographic techniques to study their microstructure alterations. To reveal grain boundaries, as-prepared samples were etched using Keller reagent which has the composition of 2.5 ml HNO_3, 1.5 ml HCl, 1.0 ml HF, and 95 ml H_2O out of 100 ml. A Field Emission Scanning Electron Microscopy (FEG-SEM) Model: FEI Nova

Nano 430 FEG-SEM ultra-high-resolution equipped with an Energy Dispersive X-ray Spectroscopy (EDS) instrument was used for microstructure observations and micro-chemical analysis at an accelerating voltage of 15 keV. Electron Probe Micro Analyser (EPMA) Model: JEOL JXA-8230 equipped with Wavelength Dispersive X-ray Spectroscopy (WDS) was also used to carry out precise chemical mapping of the distribution of Iron in the Aluminum matrix. In addition to these, the X-Ray Diffractometer (XRD) Model: Rigaku, ULTIMA-IV was used to study the qualitative phase analysis of all the samples through the x-ay diffraction method.

2.3 Microhardness Analysis

The Microhardness test was carried out using LEICA VMHTAUTO Microhardness Tester, Germany. In these measurements, the different loads of 10 kg and 15 kg were used to optimize the indentation on the soft matrix and hard intermetallic phases, respectively. Along with this, the dwell time of 15 s for each hardness indentation was used for both soft as well as hard phases.

3 Results and Discussion

3.1 Microstructure and SEM–EDS Analysis

3.1.1 As-Cast Alloy

The optical microstructure of as-cast alloy confirmed that the matrix comprising the typical dendritic structure along with uniformly distributed thick and long needle-shaped precipitates as shown in Fig. 1. The comprehensive SEM–EDS micro-analysis of these precipitates was carried out to evaluate the compositional variation; it is shown in Fig. 2. Further, based on the result of comprehensive micro-analysis, phases formed in the as-cast alloy were estimated. The SEM–EDS micro-analysis of various points along with composition-based stoichiometry of possible phases was shown in Tables 1 and 2. Mostly $Al_{13}Fe_4$ was identified which is a well-known brittle and the very hard second phase in the Iron-rich Aluminum matrix.

3.1.2 Suction Cast Alloy

The as-cast alloy was again re-melted and subsequently suction cast into strips of 2 mm in an argon atmosphere. Herein the optical microstructure of the suction cast sample, a fine needle-shaped intermetallic phase can be seen with additional rosette-like precipitates in the aluminum matrix which is a different phase morphology in

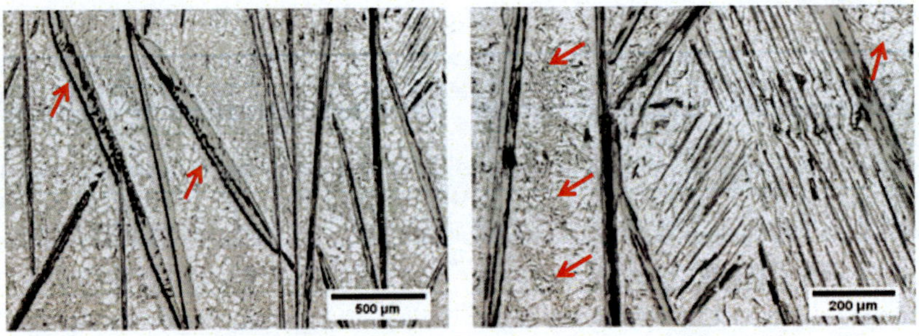

(a) (b)

Fig. 1 Optical microstructure of **a** as-cast alloy showing very coarse plate or needle shape second phase formation and **b** shows the fine dendritic microstructure in the matrix

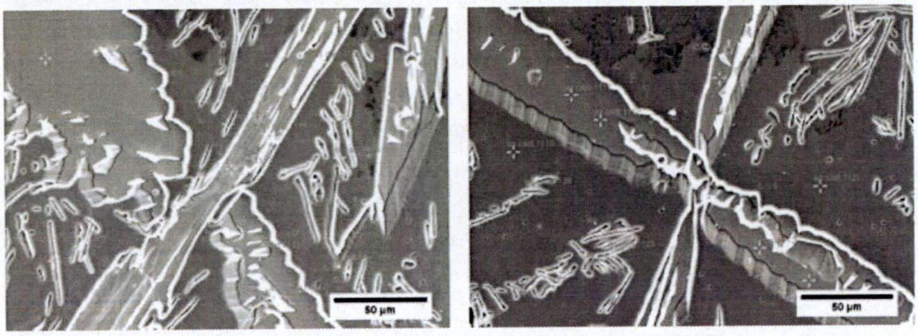

(a) (b)

Fig. 2 Shows SEM–EDS micro-analysis at various points of intermetallic phases in the as-cast alloy

Table 1 Phase identification from composition analysis

Phase	Al Wt%	Fe Wt%
$Al_{13}Fe_4$	61.09	38.91
Al_6Fe	74.35	25.65
Al_5Fe_2	54.71	45.29
$Al_{0.9}Fe_{0.1}$	81.30	18.70

Table 2 Phase estimated from element composition stoichiometry

Precipitate ID	Phases	Precipitate ID	Phases
1109	$Al_{13}Fe_4$	1117	$Al_{13}Fe_4$
1111	$Al_{0.9}Fe_{0.1}$	1118	$Al_{13}Fe_4$
1113	$Al_{13}Fe_4$	1119	$Al_{13}Fe_4$
1115	$Al_{0.9}Fe_{0.1}$	1120	$Al_{13}Fe_4$

Fig. 3 Optical microstructure of suction cast alloy showing **a** general microstructure of suction cast alloy and **b** at high magnification of the yellow circled region of "Fig. 3a."

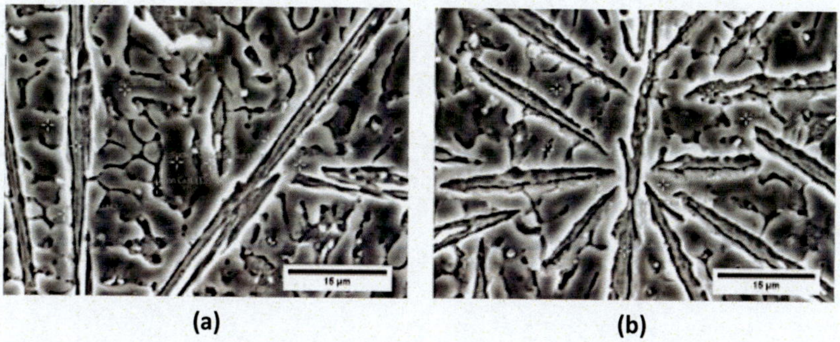

Fig. 4 SEM–EDS point analysis of various intermetallic phases in suction cast alloys

suction cast alloy as shown in Fig. 3b. Subsequently, the SEM–EDS micro-elemental analyses of the rosette-like precipitates were done as shown in Fig. 4a–b. This consequence mostly appeared as an $Al_{0.9}Fe_{0.1}$ type of phase.

3.1.3 Suction Cast and Annealed Alloy

The suction cast strips were subjected to annealing treatment under ultra-fast heating and cooling rates. The operating parameters in all annealing simulation experiments were as follows—the heating rate at 10°/s, cooling rate at 100°/s, and holding time was kept for 1 min. Together with these, the temperatures chosen for the same experiments were 800 and 825 °C. The optical microstructures observed in the suction cast samples after annealing at a high temperature are shown in Fig. 5. As per the microstructures in the picture, it can be noticed that with increasing temperature the needle-shaped intermetallics and rosette-like precipitates were fragmented down to smaller magnitudes. In addition to this, SEM–EDS micro-elemental analysis of

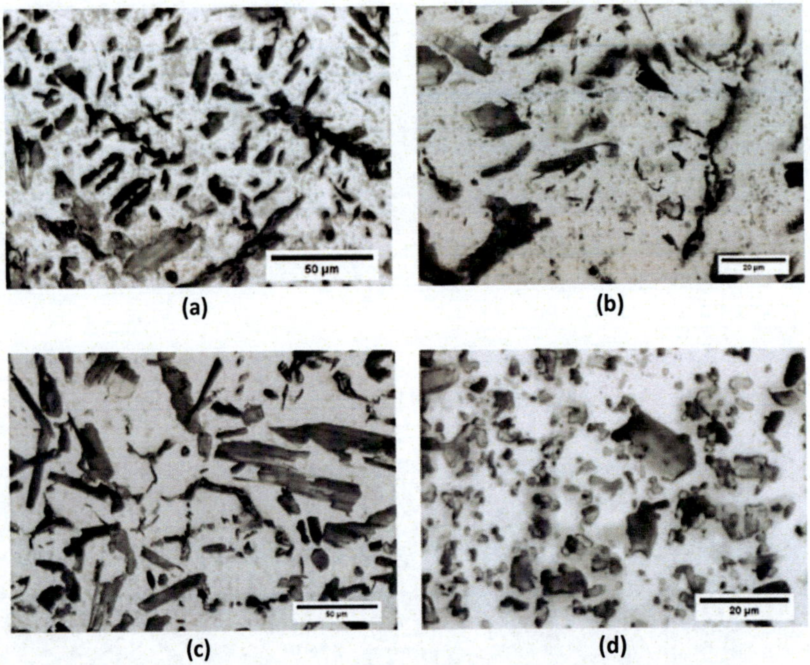

Fig. 5 Microstructures of suction cast annealed sample **a** & **b** at 800 °C and **c** & **d** at 825 °C for holding time of one minute each

annealed samples was also carried out to understand the composition variation of the precipitates as shown in Fig. 6. Based on the composition variation $Al_{0.9}Fe_{0.1}$, Al_6Fe, $Al_{13}Fe_4$, and Al_5Fe_2 phases were estimated and enlisted in Table 3.

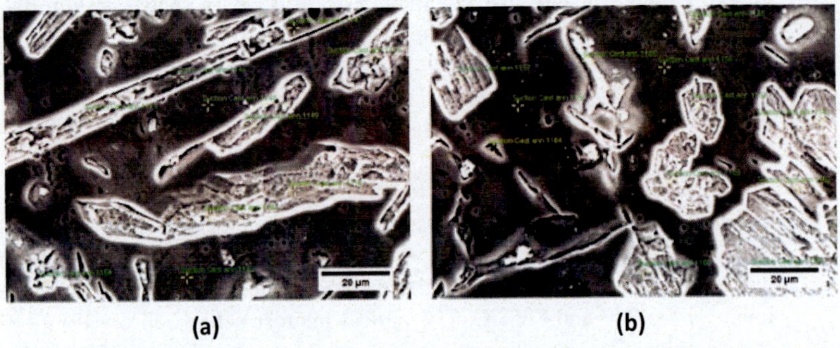

Fig. 6 SEM–EDS micro-elemental analysis at various points of intermetallic phases in **a** As-suction cast alloy and **b** 825 °C annealed alloy

Table 3 Typical SEM-EDAX point analysis of suction cast alloy

Precipitate ID	Phase	Precipitate ID	Phase
1128	$Al_{0.9}Fe_{0.1}$	1138	$Al_{0.9}Fe_{0.1}$
1133	$Al_{13}Fe_4$	1139	$Al_{0.9}Fe_{0.1}$

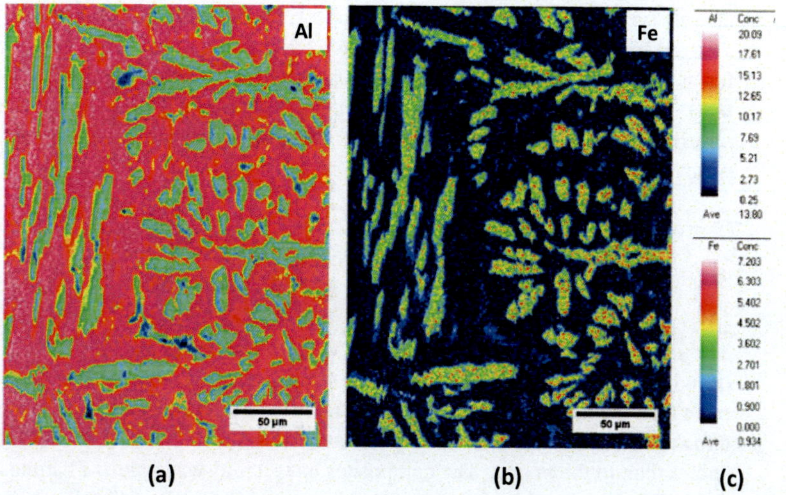

Fig. 7 A typical EPMA-WDS elemental mapping for **a** Al and **b** Fe of Al–Fe annealed alloy at 825 °C. **c** Shows the concentration scale of Al and Fe as in Fig. 7a, b

Further, the specimens were investigated for selective precipitates micro-analysis through EPMA-WDS as shown in Fig. 7. This examination was comprised of area mapping of Aluminum and Iron elements to validate the results obtained from SEM–EDS analysis.

It can be seen from the elemental mapping of Aluminum in Fig. 7a that the main matrix and the precipitates had a good concentration of Aluminum by mass beside the inter-phase was also found rich in Aluminum concentration. Again the elemental mapping of Iron in Fig. 7b shows that Iron mostly presents in intermetallic where red and pink locations indicate Iron-rich areas. The concentration scale for EPMA-WDS analysis of Aluminum and Iron is given in Fig. 7c. It can be seen that the concentrations were varying in all the needle intermetallic phases. It was due to the different weight percent of the phases which include Al_6Fe, $Al_{13}Fe_4$, Al_5Fe_2, and $Al_{0.9}Fe_{0.1}$ type of phases.

SEM analysis of as-cast alloy shows the predominant precipitate phase as $Al_{13}Fe_4$. The rosette type structures mostly constituted of Al–Fe with some of the Al_3Fe particles. On the other hand, unique observations came from the analysis of suction cast annealed sample at 850 °C in which the phases like Al_6Fe, Al_5Fe_2, and $Al_{13}Fe_4$ were in majority.

3.2 XRD Analysis

The XRD diffractogram of as-cast and subsequently suction cast alloys are shown in Fig. 8. The as-cast alloy showed the presence of Al_3Fe as the second phase along with the $Al_{0.9}Fe_{0.1}$ phase present in a scant amount. Alike as-cast, suction cast alloy also showed the formation of similar phases but the $Al_{0.9}Fe_{0.1}$ phase was found in a larger fraction to the Al_3Fe phase. However, the very different nature of phase formation was experienced in the suction cast alloy after a rapid annealing experiment. Consequently, the formation of AlFe, Al_6Fe, $Al_{13}Fe_4$, and Al_5Fe_2 phases was observed due to ultra-fast heating and cooling annealing processes. The same phases were also validated from SEM-EDS microanalysis (Tables 3, 4 and 5).

3.3 Microhardness

The microhardness was measured on the Vickers scale by using 10 kg of load with 15 s of indention time. These measurements were carried out to characterize the phases other than the matrix. Further, the microhardness analysis showed that in suction cast alloy the matrix was softer than precipitate viz. the precipitates had double hardness values than to the matrix. The composites have a mid-way value of hardness around 105 HV. Subsequently, it was found that the matrix of the annealed alloy had the same hardness values as that of the suction cast matrix. In the annealed alloy, however, the second phase was showing an incredible hardness value of 1211HV (11.87 GPa). Some typical indentation-marked micrographs are shown in Fig. 9.

3.4 Discussion

The present research was an alternative approach for producing Al-10Fe wt.%, comprised of a composite microstructure of alpha-Al and fine $Al_{13}Fe_4$ particles via a combination of processes such as suction casting followed by annealing at rapid heating and cooling cycles. A distinct microstructure refinement was attained during solidification employing suction casting. It can be noted that the second phase plate thickness and its size reduced drastically compared to as-cast alloy as shown in Fig. 3. In addition to the second phase $Al_{13}Fe_4$ microstructure refinement, a new morphology, alike to a star shape, was observed as shown in Fig. 3b. The morphological change from plate to star-like shape was reported to take place due to a faster cooling rate (14°/s) during the solidification process [19].

Herein, the estimated cooling rates during the solidification of the suction casting process were of the order of 10^2–10^3 °/s, which was very high compared to the reported cooling rate requirement. As a result in the second processing stage of

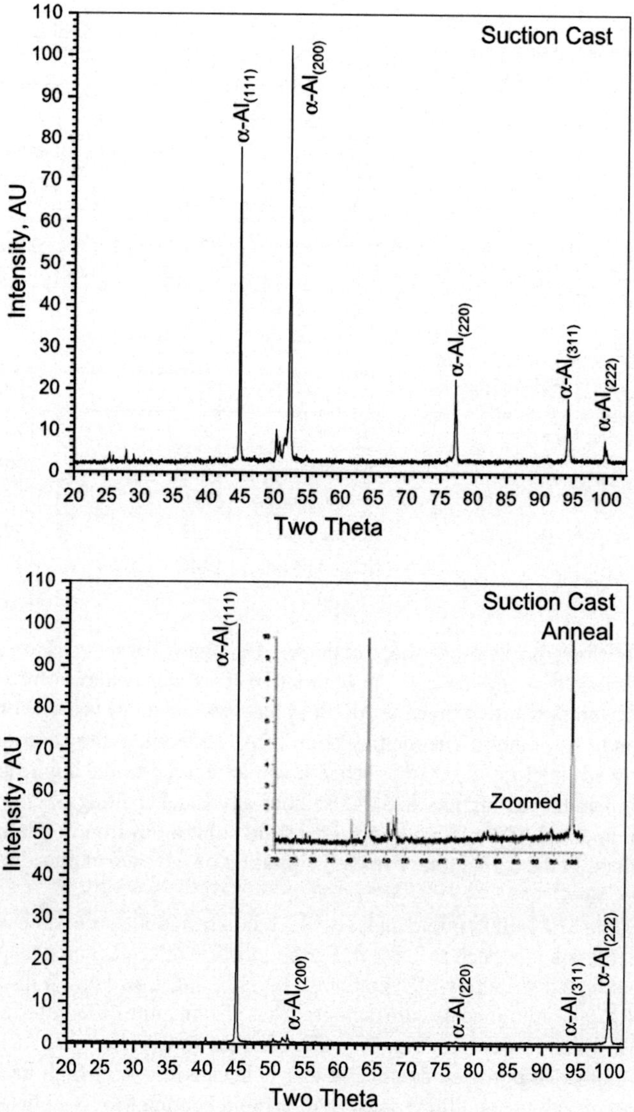

Fig. 8 XRD analysis of the suction cast and subsequently annealed alloy depicted the presence of the $Al_{13}Fe_4$ phase other than α-Al

Table 4 Typical SEM–EDS micro-elemental analysis of suction cast and annealed alloy

Precipitate ID	Phase	Precipitate ID	Phase
1145	$Al_{0.9}Fe_{0.1}$	1158	Al_5Fe_2
1146	Al_6Fe	1159	$Al_{13}Fe_4$
1149	$Al_{13}Fe_4$	1160	Al_5Fe_2
1150	$Al_{13}Fe_4$	1161	Al_5Fe_2
1151	Al_6Fe	1162	$Al_{0.9}Fe_{0.1}$
1153	Al_6Fe	1163	Al_5Fe_2
1157	$Al_{13}Fe_4$	1164	Al_6Fe

Table 5 Microhardness of the matrix in HV of the precipitate of as suction cast and after annealing

Site	As-suction cast, HV	After annealing, HV
Matrix	71	77
Second phase	141	538
Second phase	156	623
Second phase	139	405
Second phase	146	1171

annealing, the plate-like shape of the $Al_{13}Fe_4$ phase was reduced in size and uniformly distributed as fine particles in the matrix, This was achieved by annealing at very high temperatures where the $Al_{13}Fe_4$ phase was unstable because of the solubility of Iron in Aluminum. The melting point of Al-10Fe wt.% alloy according to the phase diagram is close to 900 °C, hence it can be agreed to the application of annealing temperature as high as 825 °C without any local melting. As a result of the fast cooling rate, close to a homogenous solid solution of Iron in Aluminum at 825 °C, helped in the formation of fine precipitation of the second phase which is shown in Fig. 5a–d.

The SEM–EDS microanalysis which confirmed the second phase such as Al_6Fe, Al_5Fe_2 are tabulated in Table 4. Further, the EPMA-WDS area mapping showed the chemical distribution of Aluminum, Fig. 7a, and Iron Fig. 7b in the matrix. In the XRD analysis, annealed diffractogram has shown quite different nature than a suction cast which is shown in Fig. 8. In annealed the (hkl) reflections such as (200), (220), (311) had suppressed drastically that is because of very high heating and cooling rate annealing conditions. Importantly, rapid heating to very high temperatures, near to the melting point of the alloy, followed by rapid cooling helped in reducing the size of precipitates and abolition of the dendritic microstructure of the alloy. Finally, by controlling the process parameters, it was possible to produce an in-situ super hard Aluminum Iron alloy dispersed with fine hard particles and hence enriched the hardness of the system.

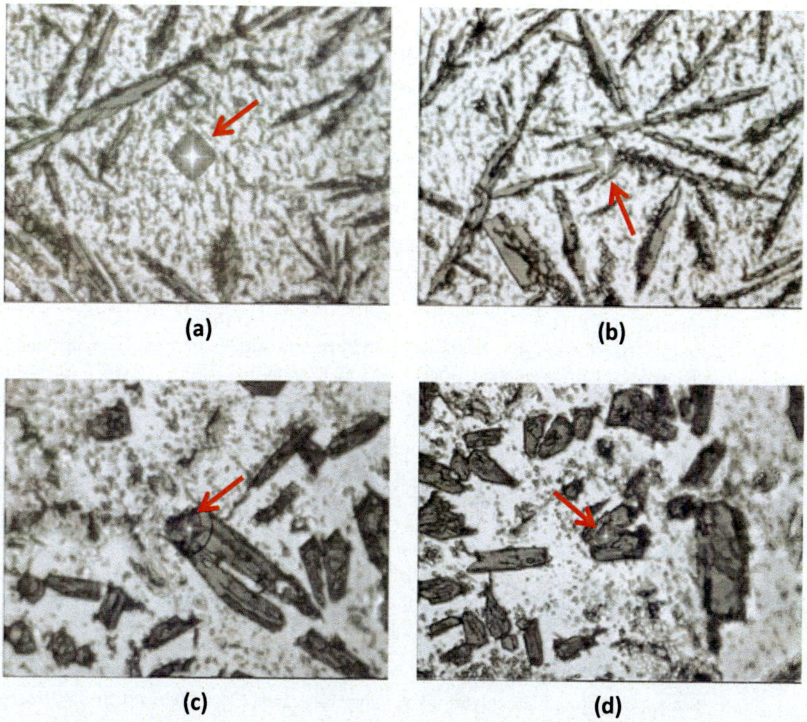

Fig. 9 Illustration of the points of indentations in suction cast sample **a** matrix **b** composite area **c** precipitate **d** precipitate of the annealed sample

4 Conclusions

The current investigation was focused to cast Al-10Fe wt.% alloy and refine the undesirable microstructure, for instance, very long and thick plate-like shape of the $Al_{13}Fe_4$ phase explicitly formed during the casting of the alloy. This undesirable phase has very high-stress concentrations which deteriorate the mechanical properties of the alloy. Hence the as-cast alloy was subsequently suction cast having a cooling rate of the order of 10^2–10^3 °/s which reduced the size of the plate-like structure down to a certain extent. Further, the suction cast alloy was gone through rapid annealing processes at very high temperature viz. above 800°/C under very high heating and cooling rate of 10 and 100°/s, respectively. As a result of this, the $Al_{13}Fe_4$ phase was reduced significantly in size and distributed uniformly in the matrix. The consequences of this study indicate a possible process route of producing Al-10Fe wt.% alloy with fine hard and well dispersed second intermetallic phases. Interestingly, the present annealing process showed a variety of second phase formation of the type Al_6Fe, $Al_{13}Fe_4$, Al_5Fe_2, and $Al_{0.9}Fe_{0.1}$ and hardness levels had increased to as high as 11.87 GPa.

Acknowledgements The authors express sincere thanks to the Director, CSIR-National Metallurgical Laboratory, Jamshedpur for his kind support and permission to publish this work.

References

1. Lendvai A (1986) Phase diagram of the Al-Fe system up to 45 mass % iron. J Mater Sci Lett 5:1219–1220
2. Kattner UR, Burton BP (1992) ASM handbook. ALLOY phase diagrams: ASM. Materials Park, OH, pp 2–44
3. Belov NA, Aksenov AA, Eskin DG (2002) Iron in Aluminum alloys. Impurity and alloying element, 1st ed. Taylor and Francis, London, U.K, pp 1–7
4. Hatch JE (1984) Aluminum: properties and physical metallurgy, 1st ed. ASM, Materials Park, OH, pp 25–57
5. Tcherdyntsev VV, Kaloshkin SD, Gunderov DV, Afonina EA, Brodova IG, Stolyarov VV et al (2004) Phase composition and microhardness of rapidly quenched Al–Fe alloys after high pressure torsion deformation. Mater Sci Eng A 375:888–893
6. Mukhopadhyay DK, Suryanarayana C, Froes FH (1995) Structural evolution in mechanically alloyed Al-Fe powders. Metall Mater Trans A 26:1939–1946
7. Cardellini F, Contini V, Mazzone G (1996) Solid-state reactions in the Al-Fe system induced by ball milling of elemental powders. J Mater Sci 31:4175–4180
8. Cardellini F, Contini V, Gupta R, Mazzone G, Montone A, Perin A, Principi G (1998) Microstructural evolution of Al-Fe powder mixture during high-energy ball milling. J Mater Sci 33:2519–2527
9. Senkov ON, Froes FH, Stolyarov VV, Valiev RZ, Liu J (1998) Microstructre and microhardness of an Al-Fe alloy subjected to sever plastic deformation and aging. Nanostruct Mater 10:691–98
10. Stolyarov VV, Lapovok R, Brodova IG, Thomson PF (2003) Ultrafine-grained Al-5 wt. % Fe alloy processed by ECAP with backpressure. Mater Sci Eng A 357:159–67
11. Senkov ON, Froes FH, Stolyarov VV, Valiev RZ, Liu J (1998) Microstructure of aluminum-iron alloys subjected to severe plastic deformation. Scr Mater 38:1511–1516
12. Li L, Zhang YD, Esling C, Jiang HX, Zhao ZH, Zuo YB, Cui JZ (2012) Influence of a high magnetic field on the precipitation behavior of the primary Al3Fe phase during the solidification of a hypereutectic Al-3.31 wt% Fe alloy. J Cryst Growth 339:61–69
13. Zhang YH, Liu YC, Han YJ, Wei C, Gao ZM (2009) The role of cooling rate in the microstructure of Al-Fe-Si alloy with high Fe and Si contents. J Alloy Compd 473:442–445
14. Moldovan P, Popescu G, Miculescu F (2004) Microscopic study regarding the microstructure evolution of the 8006 alloy in the plastic deformation process. J Mater Process Technol 153:408–415
15. Sun SB, Zheng LJ, Liu JH, Zhang H (2017) Selective laser melting of an Al-Fe-V-Si alloy: microstructural evolution and thermal stability. J Mater Sci Technol 33:389–396
16. Moritz DH, Schroers J, Herlach DM, Grushko B, Urban K (1998) Undercooling and solidification behavior of melts of the quasicrystal-forming alloys Al-Cu-Fe and Al-Cu-Co. Acta Mater 46:1601–1615
17. Nayak SS, Murty BS, Pabi SK (2008) Structure of nanocomposites of Al-Fe alloys prepared by mechanical alloying and rapid solidification processing. Bull Mater Sci 31:449–454
18. Dorin T, Stanford N, Birbilis N, Gupta RK (2015) Influence of cooling rate on the microstructure and corrosion behavior of Al-Fe alloys. Corros Sci 100:396–403
19. Sahoo KL, Das S, Murthy BS (2004) Formation of quasi-crystalline related intermetallic compounds in conventionally cast Al-Fe-V-Si alloy. J Non-Crystalline Solids 334 & 335:29–32
20. Sasaki H, Kita K, Nagahora J, Inoue A (2001) Nanostructures and mechanical properties of Bulk Al-Fe alloys prepared by electron-beam deposition. Matter Trans 42:1561–1565

Novel Approach on Energy Harvesting in Carbon Nanotube-Reinforced Composite Structure

S. Gopikumar, S. Uma, K. Jayakrishna, and U. Muthuraman

1 Introduction

The application of shear and its corresponding change in the deformation of beams and its assessments are projected by Ressiner (1945) and Mindlin (1959). Innovation in the field of change in order of deformation formed by beams as stages for different materials are evaluated by Noor and Burton [5]. In the past three decades applications of HSDT in the area of different composite materials has been drastically raising for its accuracy, less investment, and simplicity in calculation. Even in current research status no single theory has been proposed as the only root cause for evaluating the bend, shake, and collapse. The kinematics involved in HSDT is best explained by Murthy and Reddy which is still the most excellent preference [1, 2, 4, 6]. The energy type associated with this examination depends on active vibrations which are advanced through as basic and machine vibrations that are completed by methods for human action. The transformation components supporting the cycle are electromagnetic, piezoelectric, electrostatics. Active fuel sources can be gotten from human development, auxiliary, and hardware vibrations. Mechanical energy can be changed to power by misusing the mechanical strain or relative development in the transducer. In mechanical strain, the energy from basic mishappening can be extricated by utilizing dynamic materials, for example, piezoelectric or magnetostrictive.

S. Gopikumar (✉)
SCAD College of Engineering and Technology, Tirunelveli, India
e-mail: gopikumar@scadengineering.ac.in

S. Uma
National Institute of Technology, Tiruchirapalli, India

K. Jayakrishna
Vellore Institute of Technology, Vellore, India

U. Muthuraman
Francis Xavier Engineering College, Tirunelveli, India

R. Prasad et al. (eds.), *Advancement in Materials Processing Technology*,
Springer Proceedings in Materials 12,
https://doi.org/10.1007/978-981-16-3297-6_13

Electrostatic force generators depend on capacitance change of variable capacitors. Electrical force is produced from capacitance change between the two equal plates, which are electrically secluded from one another by vacuum, air, oil, or encasing. Piezoelectric impact was first found by Jacques Curie and Pierre in 1880. They found that specific materials can create an electrical potential when exposed to mechanical strain. The electromagnetic transducer deals with the rule of Faraday's law of electromagnetic acceptance, which expresses that the change in attractive motion produces a voltage in the loop. This generated voltage has relatively higher moment rapidity and the progresses are comparatively eye-catching. The electromagnetic enlistment can be acknowledged by either a moving magnet with a fixed curl or a moving loop with a fixed magnet.

1.1 Objective

To plan a brilliant material conveying a sensor property that covers bigger territory without leaving hints of unavailable structures.

To assess the temperature and weight utilizing the planned brilliant material then reaping the energy which are squandered in open structures.

2 Materials and Methods

In this experimentation the piezoelectric components PolyVinyli Dene Fluoride (PVDF) and Lead Zirconate Titanate (LZT) are used as energy harvesting device which are strong polymers, non-reactive with chemicals, and has strong mechanical properties and easily attachable to the surface of concrete (Tables 1 and 2).

The reinforcement of carbon nanotube concrete framed with sole string of 14 mm steel along with 0.18 m depth and 0.075 m width are used for experimentation, where its efficient span comprises 1.9 m. The load is subjected to fall on top surface where the harvesting device is enclosed in the bottom (Fig. 1).

3 Result and Discussion

The utilization of energy collector offers practical and alluring arrangement. One among the identified three separate energy collecting innovations, it was chosen to use piezoelectric material to shape the premise of the energy reaping gadgets. The explanation of this that the materials can be appended either as a cantilever-based gatherer or as cement fixed straight forwardly to the host structure.

Finite element analysis and investigational appraisal of the casted beam based on CNT-RCC on damage detections are carried out using energy harvesting system.

Table 1 Comparative study on energy conversion system

S. no	Techniques	Positivity	Negativity
1	Piezoelectric	The power density is higher along with higher output voltage There is no need for an external bias source The piezoelectric device may be micro- and macro-level The method of implementation is simple	Material fatigue occurs at certain stage due to bending moment
2	Electromagnetic	Accurate with low frequency There is no need for an external bias source The power density obtained is neutral The flow of current is higher and arise low impedance	It is not advisable for micro setup Need large mass displacement
3	Electrostatic	Flexible with low-frequency applications High-quality output voltage is attainable	Need initial bias voltage The power densities obtained are relatively lesser

Table 2 Properties of composite energy harvester

S. no	Property	Symbols	Value	Unit
1	Piezoelectric strain constant	d_{31}	21	pC/N
2	Dielectric permittivity	e	102	pF/m
3	Density	ρ	1769	Kg/m^3
4	Young's modulus	E	8.1	GPa
5	Material thickness	t	51	μm
6	Poisson's ratio	υ	0.17	

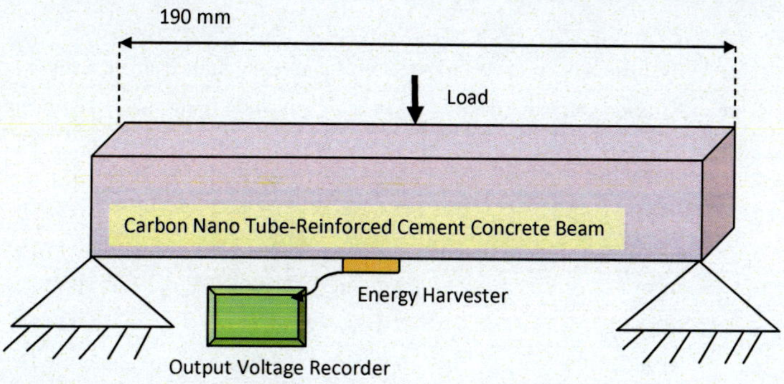

Fig. 1 Schematic view of experimental method

3.1 FEM Evaluation

The finite element analysis optimized the natural frequency of beam as 74.21 Hz (Fig. 2 and Table 3).

From the experimental outcome, it is clear that open crack has been identified corresponding to the applied load. With respect to natural frequency the transferred data are evaluated and observed to be decreasing compared to various crack depth ratio (Fig. 3 and Table 4).

The voltage observed through the energy harvester is allowed to store as DC Voltage with the support of capacitor, and nowadays it is easy to rectify the stored DC Voltage to AC. With the support of circuits the energy harvesting process may be carried forward to the entire open building by means of capturing the voltages.

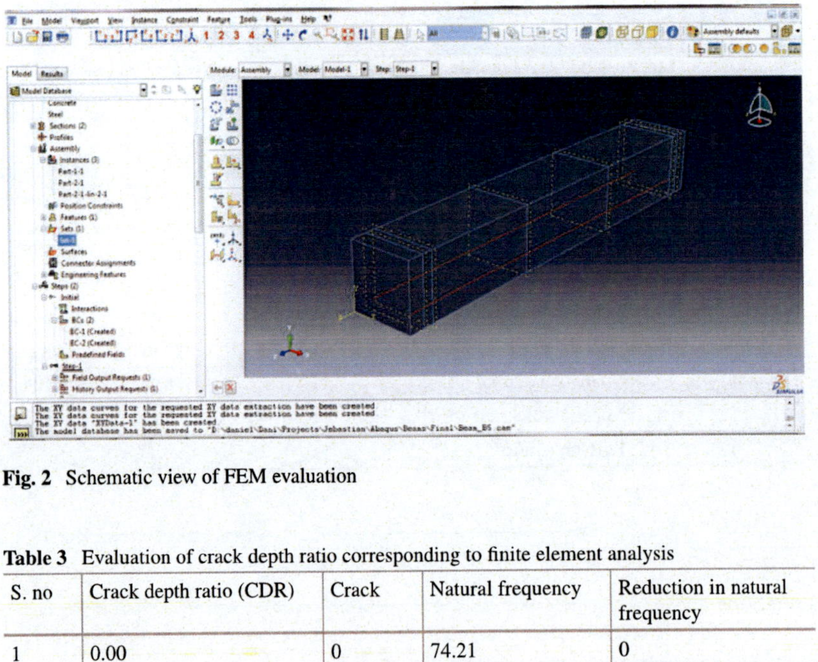

Fig. 2 Schematic view of FEM evaluation

Table 3 Evaluation of crack depth ratio corresponding to finite element analysis

S. no	Crack depth ratio (CDR)	Crack	Natural frequency	Reduction in natural frequency
1	0.00	0	74.21	0
2	0.04	7.8 mm	74.03	0.18 Hz
3	0.08	7.4 mm	73.64	0.57 Hz
4	0.12	7.2 mm	73.76	0.45 Hz
5	0.16	7.6 mm	73.7	0.51 Hz
6	0.20	7.1 mm	73.75	0.46 Hz

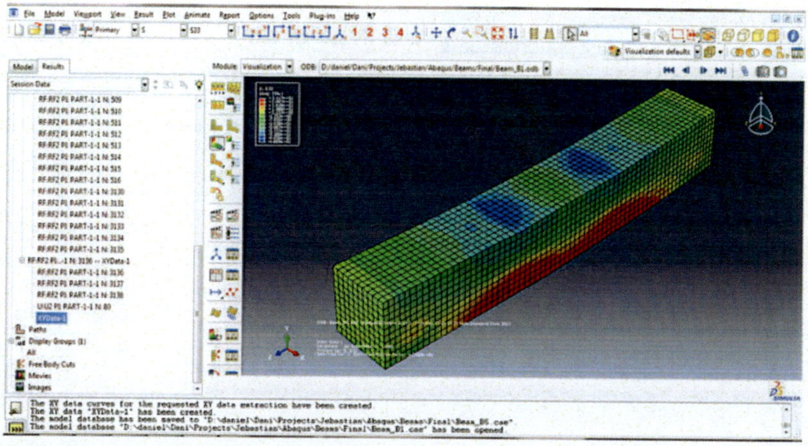

Fig. 3 Schematic view of beam after testing

Table 4 Evaluation of crack depth ratio corresponding to experimental method

S. no	Crack depth ratio (CDR)	Crack	Natural frequency	Reduction in natural frequency
1	0.00	0	74.26	0
2	0.04	7.1 mm	73.9	0.36 Hz
3	0.08	6.9 mm	73.4	0.87 Hz
4	0.12	7.4 mm	72.91	1.35 Hz
5	0.16	7.2 mm	71.75	2.51 Hz
6	0.20	6.8 mm	70.8	3.46 Hz

4 Conclusion

Structural health monitoring along with energy harvesting techniques are well experimented and the optimized values are CDR 0.20 and Natural Frequency 70.8, respectively.

References

1. Bickford WB (1982) A consistent higher order beam theory. In. Development in theoretical and applied mechanics. vol XI. University of Alabama
2. Levinson M (1980) An accurate simple theory of statics and dynamics of elastic plates. Mech Res Commun 7:340–350
3. Mindlin RD (1951) Influence of rotatory inertia and shear on flexural motion of isotropic, elastic plate. Mech J Appl Mech 18:31–38

4. Murthy VV (1981) An improved transverse shear deformation theory for laminate anisotropic plates. NASA technical paper 1903
5. Noor AK, Burton WS (1989) Assessment on shear deformation theories for multilayered composite plates. Appl Mech Rev 42(1):1–13
6. Reddy JN (1984) A refined nonlinear theory of plates with transverse shear deformation. Int J Solids Struct 20:881–896
7. Reddy JN (1984) A simple higher-order theory for laminated composite plates. J Appl Mech 51:745–752
8. Reissner E (1945) The effect of transverse shear deformation on the bending of elastic plates. J Appl Mech 12:66–77

Numarical Analysis of Mechanical Ventilation Systems

Raviranjan Kumar Singh, Priya Ranjan Mishra, Rina Sahu, and Shristi Singh

1 Introduction

Buildings and structures aim to offer protection against the wind, sun, rain and cold. Earlier, designs used to be comparatively simple and were constructed depending upon the environmental aspects of the surroundings. The interior comfort was enhanced by natural ventilation as well as thermal mass, which provide passive cooling. This showed that comfortability was prioritized for both—residence and workspace. Since humans spend their maximum time indoors, the estimation of indoor environment conditions is vital for its optimization during designing. A set of various indoor climate parameters has to be determined for proper air quality and thermal comfort of the occupants.

Mechanical ventilation systems ensure proper air quality by circulating fresh air through fans/ducts instead of depending on small cracks or holes in windows, roofs, or walls for airflow. Choosing the best strategy for the ventilation of a room depends on various features, such as the building's purpose. For a workspace, mixing ventilation and displacement ventilation are the two main principles of mechanical ventilation. Mixing ventilation features supplying air into a room at a considerably high velocity. The entrainment of room air to be mixing high degree and the temperature and contaminant concentration tends to uniform. Since the concept of mechanical ventilation has been introduced, mixing ventilation has been the most essential ventilation principle.

Cockroft and Robertson [1] found that the ventilation in a building is influenced by the steady effect of temperature, air pressures as well as the turbulence of wind.

R. K. Singh (✉)
Department of ECEN, KIT Varanashi, Varanashi 221307, UP, India

P. R. Mishra · R. Sahu · S. Singh
Metallurgical and Materials Engineering, NIT-Jamshedpur, Jharkhand 831007, India

R. Prasad et al. (eds.), *Advancement in Materials Processing Technology*, Springer Proceedings in Materials 12,
https://doi.org/10.1007/978-981-16-3297-6_14

Awbi [2] presented the results of an application created to solve 2D and 3D ventilation models, which show the results of the distribution of temperature and air velocity and corroborated with the experimented measurements. Awbi [3] examined the CO2 distribution and air movement in a naturally ventilated workspace as well as an atrium through CFD Model. Yuan [4] presented detailed experimented information about the airflow in a room using displacement ventilation system model. Computational Fluid Dynamics program was used along with RNG k-ε model to predict the indoor airflow. Li [5] derived the solution for thermal stratification as well as airflow for buoyancy-driven ventilation in a single zone building. It was further observed that the fully mixed model estimated a comparatively lower neutral level height for the buildings than the height estimated through models that considered thermal stratification. Stern et al. [6] presented a detailed strategy for validating the procedures and methodology of CFD simulations developed using a pre-developed CFD code prepared in the Indore validation system. Chen and Srebric [7] investigated that Computational Fluid Dynamics model to measure the heat transfer, airflow, and transport of chemical species for analyzing a large variety of o HVAC&R and the conditions of the indoor environment. Allocca et al. [8] studied natural ventilation as an efficient means to save energy and to enhance the air quality of the indoors in a building. Zhai [9] examined the Computational fluid dynamics for most sophisticated airflow modelling method, which estimates heat transfer, airflow, and contaminant transportation simultaneously around buildings. Stamou and Katsiris [10] used the SST k–ω based model for calculating the velocities of airflow as well as the temperature inside a standard office-room. It was developed for the combination of a suitable grid. Srebric et al. [11] found that the stimulations of computational fluid dynamics for the indoor environment estimate contaminant dispersion around human occupants and provide valuable information in fixing indoor air quality or homeland security issues. Abdilghanie et al. [12] compared turbulence modelling techniques for the simulation of low-speed indoor airflow in a room. The impact of the intensity of inlet turbulence on the flow field through the constant-coefficient large eddy simulation model with uniform mean inlet conditions at a wide range of inlet turbulence was also investigated. Chen [13] described the tools for prediction of performance of ventilation in buildings. These reviewed tools were empirical models, analytical models, small-scale experimental models, multi-zone network models, zonal models, full-scale experimental models, and Computational Fluid Dynamics models. Wang and Chen [14] study and show the advanced features for solving airflow in enclosed environment using LES and DES models. Lipska et al. [15] states the theory behind air-conditioning and thermal environment of the monumentally famous auditorium at the University of Wroclaw. Hurnik et al. [16] created a new CFD 3D benchmark with two well-defined characteristic zones in the room. They proposed the validation of the CFD results and also reported methods suitable for the benchmark data. For testing various aspects of CFD modelling, the new 3D benchmark may be beneficial for CFD field researchers, engineers, and students. It may also lead to the improvement of the accuracy of CFD prediction. Nielsen [17] discussed the CFD principle, the advancement in numerical methods, and turbulence models.

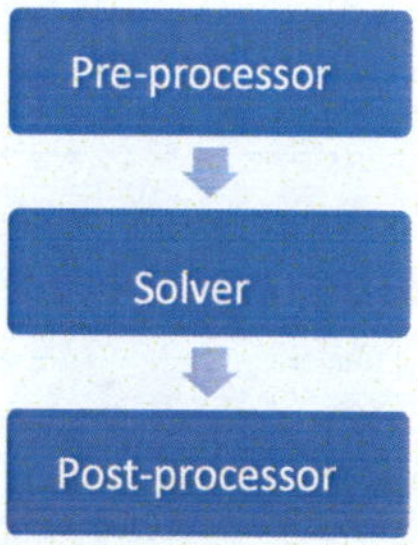

Fig. 1 Sequence of process in computational fluid dynamics

2 CFD Modelling and Analysis

2.1 Introduction

Computational fluid dynamics (CFD) is a branch of fluid mechanics that uses numerical analysis and data structures to solve and analyze problems that involve fluid flows. Computers are used to perform the calculations required to simulate the interaction of liquids and gases with surfaces defined by boundary conditions. The widespread availability of computational power at reasonable rate and also its span of application from modelling to simulation, design optimization and manufacturing had shifted the preference towards this approach of study. CFD modelling and analysis can be divided into four parts, i.e., geometry modelling, meshing, computer simulation, and data reduction. Geometry modelling and meshing constitute the pre-processor section. Computer simulation is the solver section. The data reduction is the post-processor section in which we formulate the obtained data into more conceivable forms like graph or contour plot (Fig. 1).

2.2 Geometry

Full scale 3D modelling and analysis of the room is carried out in this study. The dimensions of the room are 4.3 m × 2.85 m × 3.3 m (length × width × height). In this study for mechanical ventilation system, nine different combinations of inlet opening are modelled and analyzed. The variation in the geometry of inlet opening is discussed as follows: the three different diameters of the inlet openings are 0.3, 0.4, and 0.5 m. There are three different sizes of inlet opening at lower level (0.8 m above the floor) of the room. Similarly, there are three different sizes of inlet opening at middle level (1.65 m above the floor) and higher level (2.5 m above the floor) of the room. The size of the outlet opening is 1.4 m × 0.2 m (length × height). The outlet opening is 2.65 m above the floor. The configuration and dimension of outlet opening is same for all the nine combination of inlet opening positions and dimensions. The difference in the geometry of room in mechanical ventilation system is the diameter

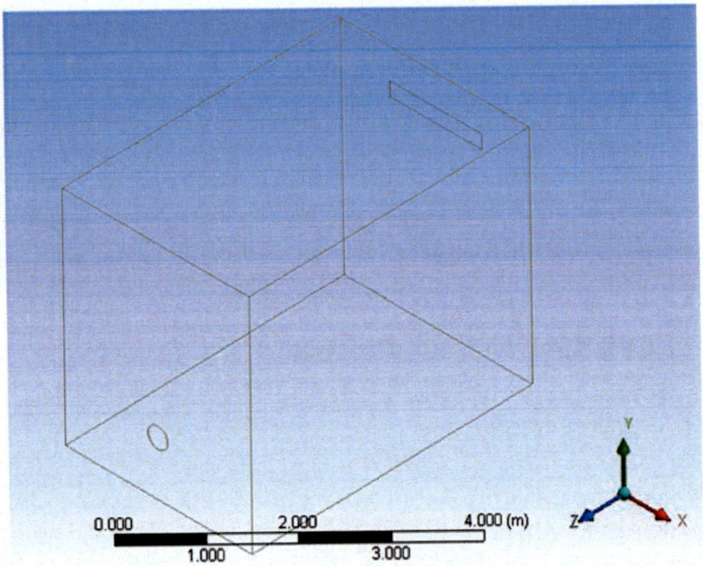

Fig. 2 Room geometry with inlet opening at low level ($d_i = 0.3$ m, $h_i = 0.8$ m)

of the inlet opening. The position of inlet is same for both natural and mechanical ventilation system. The room geometry for mechanical ventilation with diameter 0.3 m and height of inlet 0.8 m is shown in Fig. 2.

2.3 Meshing

The meshing of room geometry of mechanical ventilation system with inlet diameter 0.3 m and position of inlet 0.8 m above the floor is shown in Fig. 3.

2.4 Boundary Conditions and Solution

This study of mechanical ventilation systems is a case of transient analysis. According to the modelled geometry the direction of gravity is along negative Y-axis. Energy model is taken into consideration to calculate room temperature. In this analysis solution is done by using k-ε RNG turbulence model. The fluid inside the room is air. The initial temperature of air was taken as 313 K. The inlet temperature of air to the room was 298 K. The walls of the room are assumed to be insulated. For pressure velocity coupling SIMPLE scheme is used. Under-Relaxation factors used are shown in Table 1.

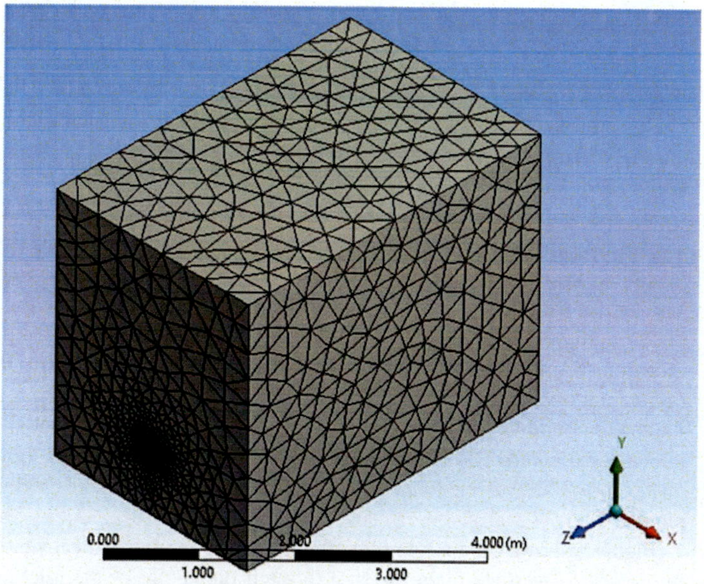

Fig. 3 Meshing of room geometry of mechanical ventilation system ($d_i = 0.3$ m, $h_i = 0.8$ m)

Table 1 Under-relaxation factors for mechanical ventilation systems

Quantity	Value
Pressure	0.3
Density	1
Body forces	1
Momentum	0.7
Turbulent kinetic energy	0.8

The mass flow rate for all the combination of inlet opening is taken constant at 0.1 kg/s. For constant mass flow rate of 0.1 kg/s the inlet velocity for three different diameters $d_i = 0.3$ m, $d_i = 0.4$ m, $d_i = 0.5$ m are calculated and shown in Table 2.

Table 2 Inlet velocities for constant mass flow rate

Inlet diameter d_i (m)	Inlet velocity v_i (m/s)
0.3	1.2
0.4	0.68
0.5	0.43

3 Results and Discussion

The analysis is done using Re-Normalisation Group (RNG) k-ε turbulence model. The calculations for natural ventilation systems are done for 5 min (300 s). After the analysis minimum temperature, maximum temperature, and average temperature of the room after 300 s are observed and stratification factor is calculated. Figure 4 shows the variation of stratification factor with inlet diameter for different height of inlet opening in mechanical ventilation system. Figure 5 shows the variation of stratification factor with height of inlet for different diameter of inlet opening in mechanical ventilation system. From Fig. 4 it can be seen that the stratification factor increases with the increase in diameter (d_i) of inlet opening. And from Fig. 5 it can be seen that stratification factor increases with the increase in the height (h_i) of the inlet opening.

The temperature and velocity contours on a vertical plane through the centre of inlet and outlet are shown in the following figures. All the contours shown are plotted on a plane parallel to YZ-plane and at a distance of 1.425 m from origin ($x = 1.425$ m plane).

Figure 6 shows the contour plot of temperature for mechanical ventilation system in the central vertical plane passing through the centre of the inlet and the outlet. This contour plot is plotted when the inlet opening is 0.8 m above the floor and the diameter of inlet opening is 0.6 m. In a similar manner Fig. 7 shows the contour plot of temperature for mechanical ventilation system when inlet opening is 0.8 m above the floor and diameter of inlet opening is 0.8 m. It can be seen from Figs. 6 and 7 that

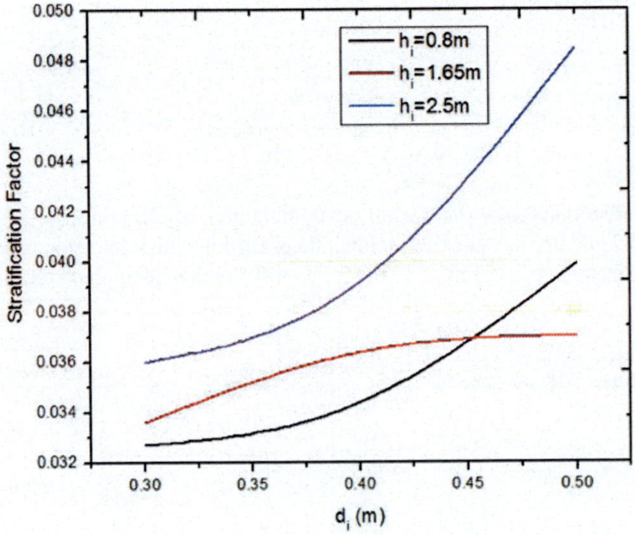

Fig. 4 Variation of stratification factor with inlet diameter for different height of inlet

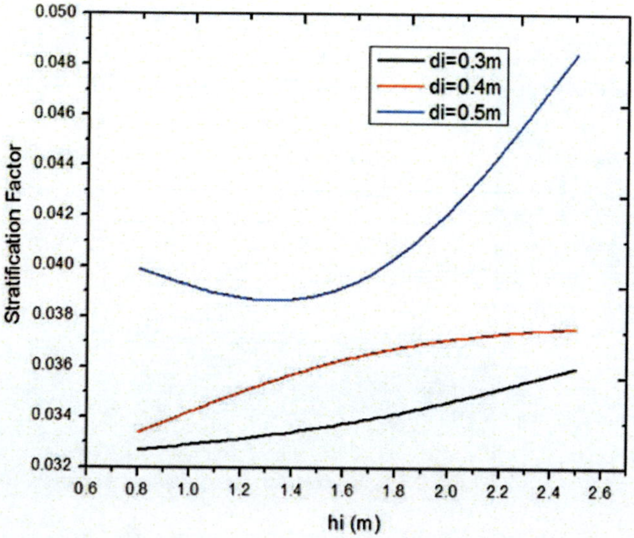

Fig. 5 Variation of stratification factor with height of inlet for different inlet diameter

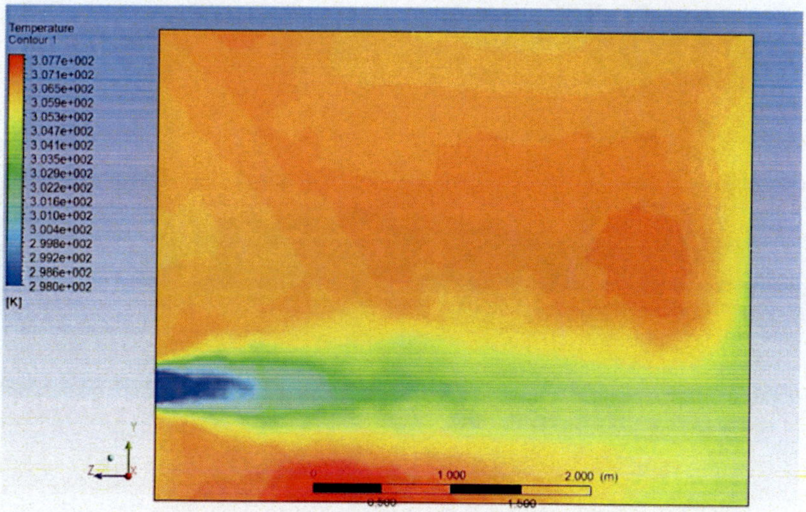

Fig. 6 Temperature contour of ventilation at 300 s for $d_i = 0.6$ m, $h_i = 0.8$ m

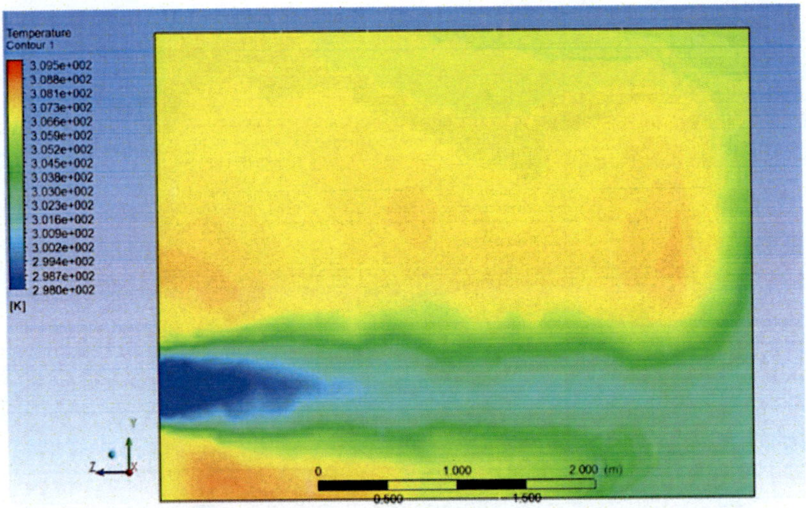

Fig. 7 Temperature contour of ventilation at 300 s for d = 0.8 m, $h_i = 0.8$ m

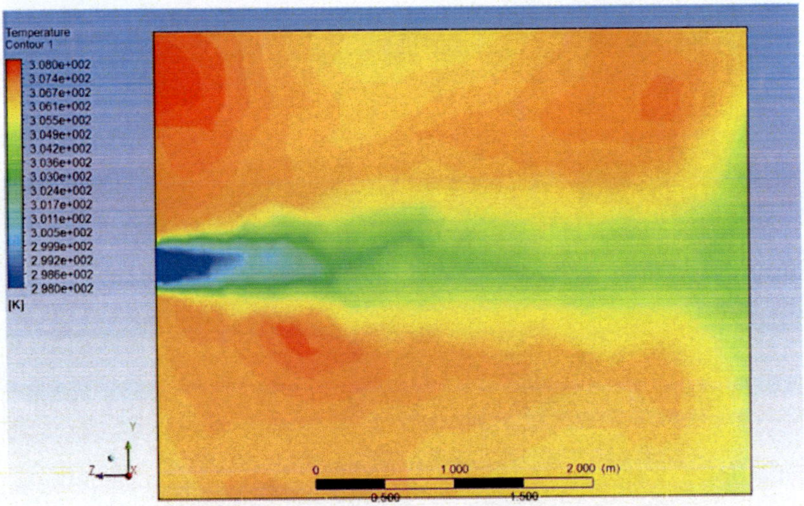

Fig. 8 Temperature contour of ventilation at 300 s for $d_i = 0.6$ m, $h_i = 1.65$ m

when diameter of inlet is increased the temperature of the occupied zone becomes less uniform. Figure 8 shows the temperature contour when inlet opening is 1.65 m above the floor and diameter of inlet is 0.6 m. Figure 9 shows the temperature contour when inlet opening is 1.65 m above the floor and the inlet diameter is 0.8 m. It can

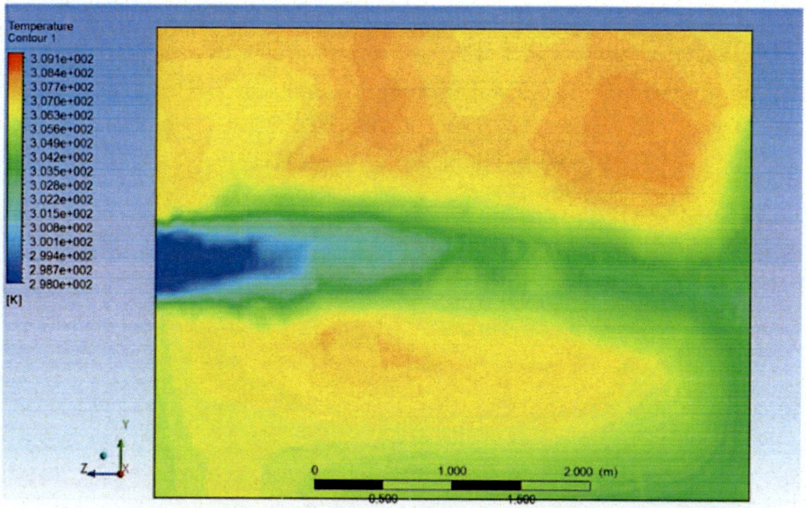

Fig. 9 Temperature contour of ventilation at 300 s for d = 0.8 m, h_i = 1.65 m

be observed for inlet opening height of 1.65 m also that the temperature contour is less uniform for larger diameter of inlet. Figures 10 and 11 shows temperature contour when inlet opening is 2.5 m above and inlet diameter are 0.6 m and 0.8 m, respectively. For inlet height of 2.5 m also, it is found that the stratification increases

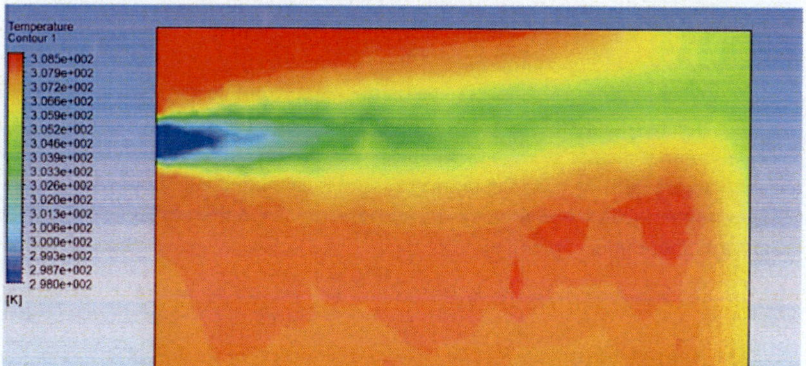

Fig. 10 Temperature contour of ventilation at 300 s for d_i = 0.6 m, h_i = 2.5 m

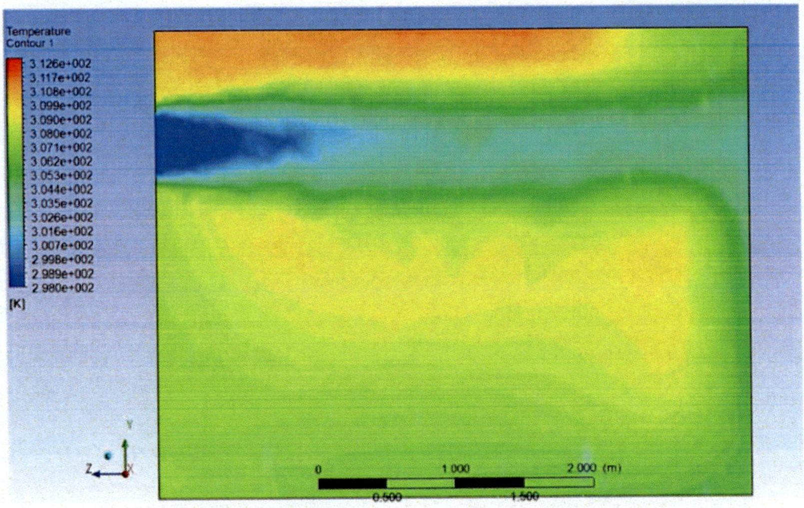

Fig. 11 Temperature contour of ventilation at 300 s for $d = 2.5$ m , $h_i = 2.5$ m

with increase of the size of the inlet opening. Hence, it has been found that at any height of inlet opening the stratification increases with the increase in inlet diameter of the opening for mechanical ventilation systems. Also, the stratification increases with the increase in the height of inlet opening for same diameter of inlet opening in mechanical ventilation system. Figure 12 shows the velocity contour plot at 300 s when inlet diameter is 0.6 m and inlet height is 0.8 m. Figure 13 shows the velocity contour plot at 300 s when inlet diameter is 0.8 m and inlet height is 0.8 m. Figure 14 shows the velocity contour plot at 300 s when inlet diameter is 0.6 m and inlet height is 1.65 m. Figure 15 shows the velocity contour plot at 300 s when inlet diameter is 0.8 m and inlet height is 1.65 m. Figure 16 shows the velocity contour plot at 300 s when inlet diameter is 0.6 m and inlet height is 2.5 m. Figure 17 shows the velocity contour at 300 s when inlet diameter is 0.8 m and inlet height is 2.5 m. From velocity contours it can be seen that at lower position of the inlet opening the movement of air is more in occupied zone.

4 Conclusions

In the present investigation, influence of positions and dimensions of inlet opening on mechanical ventilation systems with same outlet position and dimension has been analyzed. CFD modelling and analysis of mechanical ventilation system is done and the following conclusions are drawn:

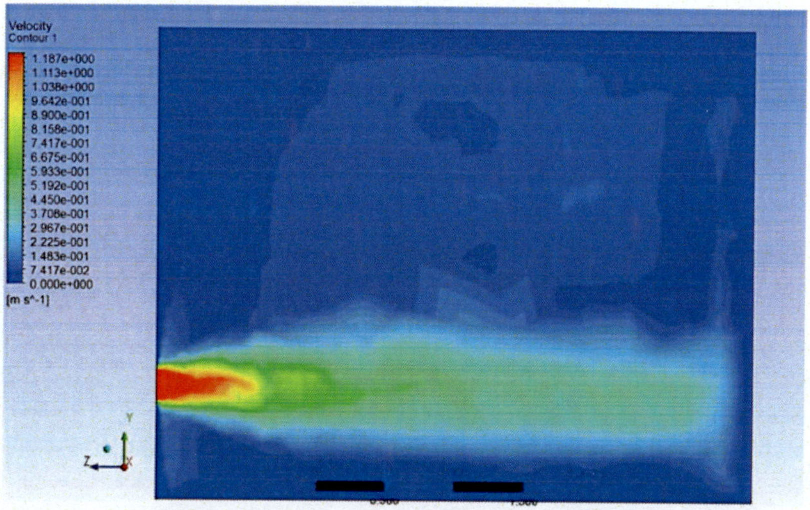

Fig. 12 Velocity contour of ventilation at 300 s for $d_i = 0.6$ m, $h_i = 0.8$ m

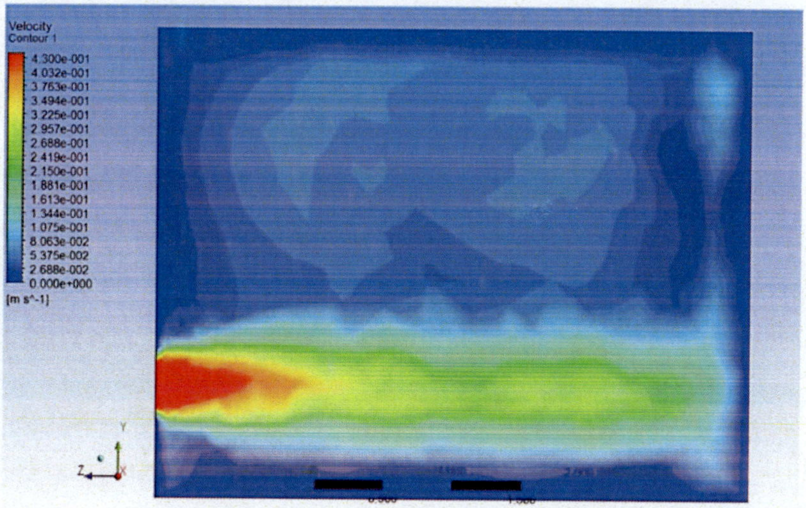

Fig. 13 Velocity contour of ventilation at 300 s for $d_i = 0.8$ m, $h_i = 0.8$ m

I. Maximum temperature and average temperature inside the room increases with increase in the size of inlet opening.
II. Stratification factor of ventilation system increases with the increase in the size of the inlet opening.

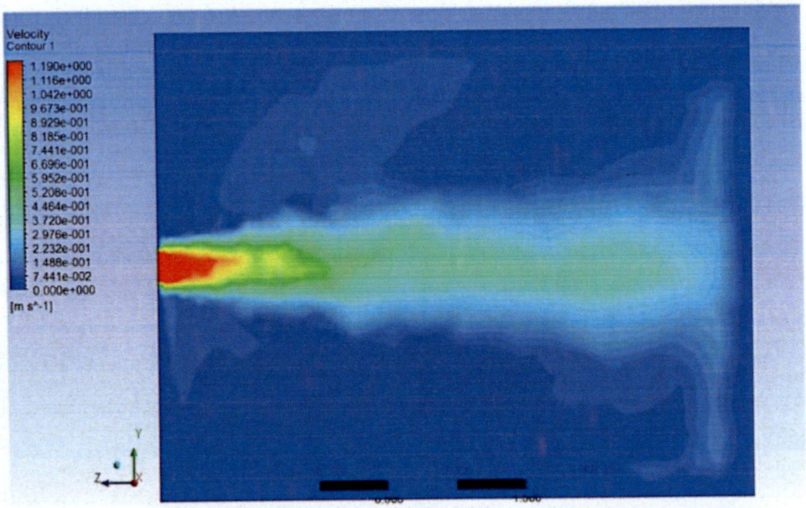

Fig. 14 Velocity contour of ventilation at 300 s for $d_i = 0.6$ m, $h_i = 1.65$ m

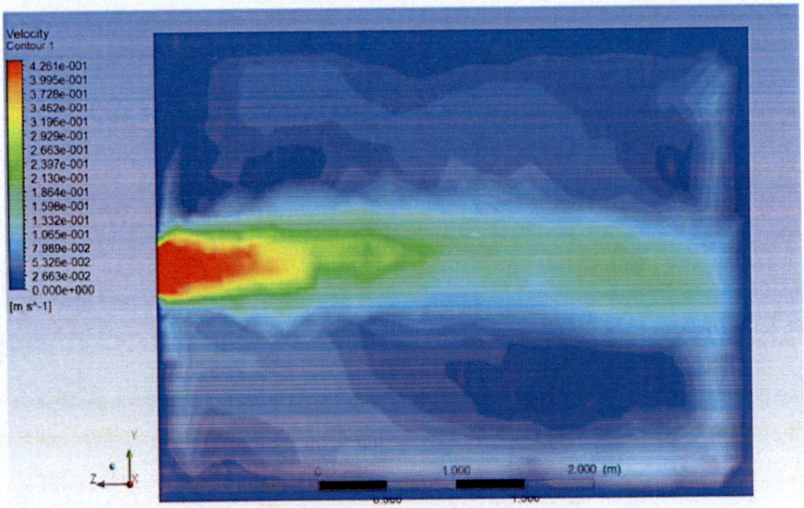

Fig. 15 Velocity contour of ventilation at 300 s for $d_i = 0.8$ m, $h_i = 1.65$ m

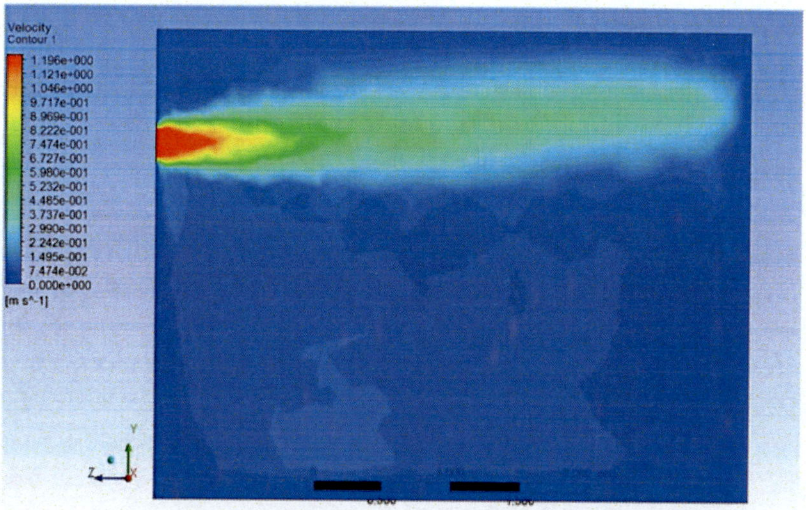

Fig. 16 Velocity contour of ventilation at 300 s for $d_i = 0.6$ m, $h_i = 2.5$ m

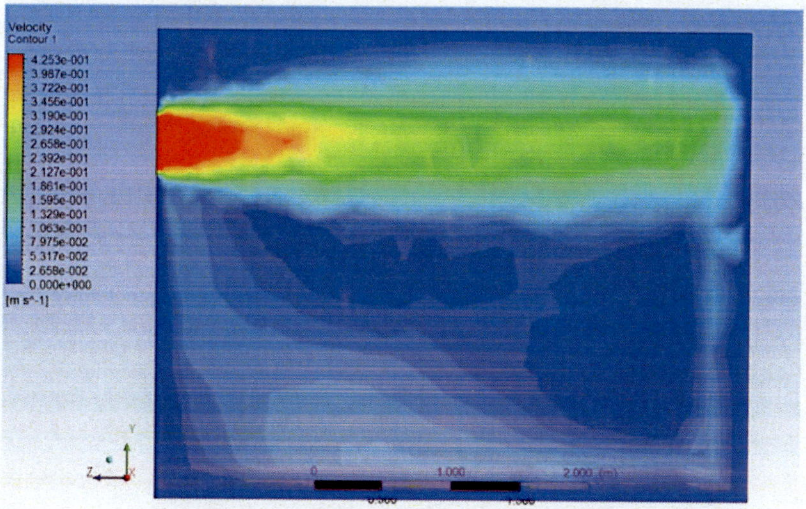

Fig. 17 Velocity contour of ventilation at 300 s for $d_i = 0.8$ m, $h_i = 2.5$ m

References

1. Cockroft JP, Robertson P (1976) Ventilation of an enclosure through a single opening. Build Environ 11(1):29–35
2. Awbi HB (1989) Application of computational fluid dynamics in room ventilation. Build Environ 24(1):73–84
3. Awbi HB (1996) Air movement in naturally-ventilated buildings. Renew Energy 8(1–4):241–247
4. Yuan X et al (1999) Measurements and computations of room airflow with displacement ventilation. Ashrae Trans 105:340
5. Li Y (2000) Buoyancy-driven natural ventilation in a thermally stratified one-zone building. Build Environ 35(3):207–214
6. Stern F et al (2001) Comprehensive approach to verification and validation of CFD simulations-Part 1: methodology and procedures. Trans Am Soc Mech Eng J Fluids Eng 123(4):793–802
7. Chen Q, Srebric J (2002) A procedure for verification, validation, and reporting of indoor environment CFD analyses. HVAC&R Res 8(2):201216
8. Allocca C, Chen Q, Glicksman LR (2003) Design analysis of singlesided natural ventilation. Energy Build 35(8):785–795
9. Zhai Z (2006) Application of computational fluid dynamics in building design: aspects and trends. Indoor Build Environ 15(4):305–313
10. Stamou A, Katsiris I (2006) Verification of a CFD model for indoor airflow and heat transfer. Build Environ 41(9):1171–1181
11. Srebric J et al (2008) CFD boundary conditions for contaminant dispersion, heat transfer and airflow simulations around human occupants in indoor environments. Build Environ 43(3):294–303
12. Abdilghanie AM, Collins LR, Caughey DA (2009) Comparison of turbulence modeling strategies for indoor flows. J Fluids Eng 131(5):051402
13. Chen Q (2009) Ventilation performance prediction for buildings: a method overview and recent applications. Build Environ 44(4):848–858
14. Wang M, Chen Q (2009) Assessment of various turbulence models for transitional flows in an enclosed environment (RP-1271). HVAC&R Res 15(6):1099–1119
15. Lipska B et al (2012) The improvement of thermal comfort and air quality in the historic assembly hall of a university. Indoor Built Environ 21(2):332347
16. Hurnik M, Blaszczok M, Popiolek Z (2015) Air distribution measurement in a room with a sidewall jet: a 3D benchmark test for CFD validation. Build Environ 93:319–330
17. Nielsen PV (2015) Fifty years of CFD for room air distribution. Build Environ 91:78–90
18. Tu J, Yeoh GH, Liu C (2012) Computational fluid dynamics: a practical approach. Butterworth-Heinemann
19. Patankar S (1980) Numerical heat transfer and fluid flow. CRC Press
20. Versteeg HK, Malalasekara W (1995) An introduction to computational fluid dynamics: the finite volume method. Longman Scientific & Technical
21. ANSYS Fluent User Guide (2014) 16th Release. ANSYS Inc., Canonsborg, Pennsylvania

Machine Learning in Drilling of GFRP Composite Using ANN

Pawan Kumar

1 Introduction

In contrast to alloys, composite materials are a combination of two or more constituents, each retaining its properties. *Fiber* and *matrix* are the two main constituents of composite material. The fibers are a hard material and can be used in a continuous or discontinuous form. The classification of fiber is given in Fig. 1. Matrices are a soft material and can be polymers, metals, or ceramics. So based on matrix composites can be classified as polymeric composites, metallic composites, and ceramic composites. Fiber-reinforced polymers (FRPs) are extremely regarded for their unique properties in comparison with metal materials in various industries, including aerospace, ships, civil engineering, and so on [4]. FRPs have fine properties such as high corrosion resistance, low strength to weight, and high deformability [2]. The fibers are the main load-carrying member, while the matrix keeps them in the desired location and orientation. Matrix also protects the fibers from environmental damages and serves as a load transfer medium between fibers. Glass fiber reinforced polymers (GFRP) and carbon fiber reinforced polymers (CFRP) are the most widely used FRP in the industry.

Nowadays, machine learning techniques have been extensively utilized over traditional methods of modeling for the prediction of different machining responses in any machining processes. There are many approaches to machine learning. Broadly these methods can be classified as *supervised* and *unsupervised* [6]. The supervised approach labels input and output data while the unsupervised approach is based on unlabeled input data [1, 9]. The traditional method for the modeling of machining systems is cumbersome and requires in-depth knowledge of complex machining behaviors of composite materials. The workflow for both modeling approaches is

P. Kumar (✉)
Department of Mechanical Engineering, National Institute of Technology Kurukshetra, Kurukshetra 136119, India

R. Prasad et al. (eds.), *Advancement in Materials Processing Technology*,
Springer Proceedings in Materials 12,
https://doi.org/10.1007/978-981-16-3297-6_15

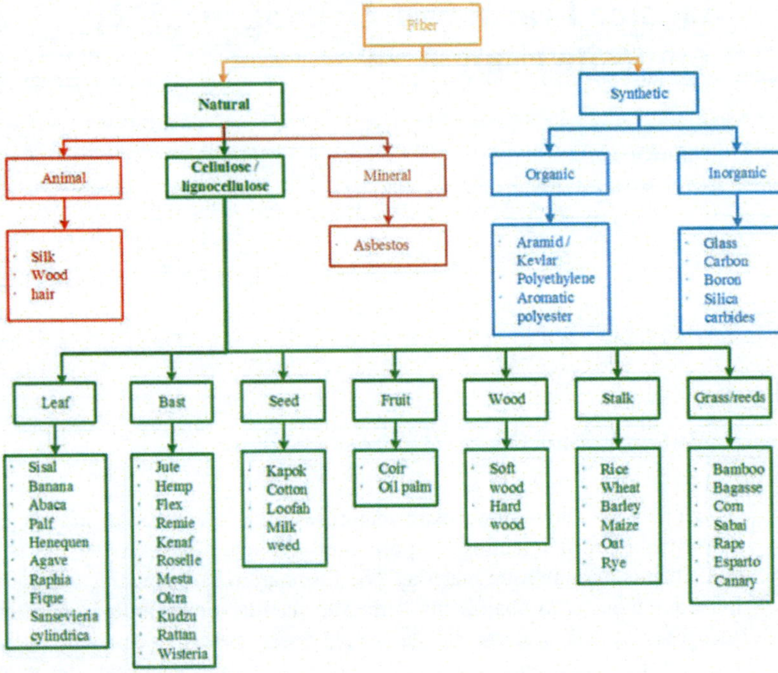

Fig. 1 Classification of natural and synthetic fiber

shown in Fig. 2. With the advancements in the information and technology sector, implementation of Industry 4.0 becomes feasible. In this study, a prediction model is developed using a supervised approach of machine learning for the prediction of MRR during drilling of GFRP using artificial neural networks (ANN) technique.

The machining behavior of composites materials is a complex phenomenon. Drilling is considered to be one of the widely used traditional techniques used in industry for composite material processing. High-speed steel (HSS) drill bits are widely used for the drilling of FRPs materials due to its easy availability and economic reason [5, 10, 11]. However, presented models in literature lack to deal with drilling operations of composite materials. Furthermore, very few utilizable information is provided regarding the prediction of the material removal rate (MRR). This research work focused on supervised learning for the prediction of MRR for the drilling of GFRP material, by using ANN.

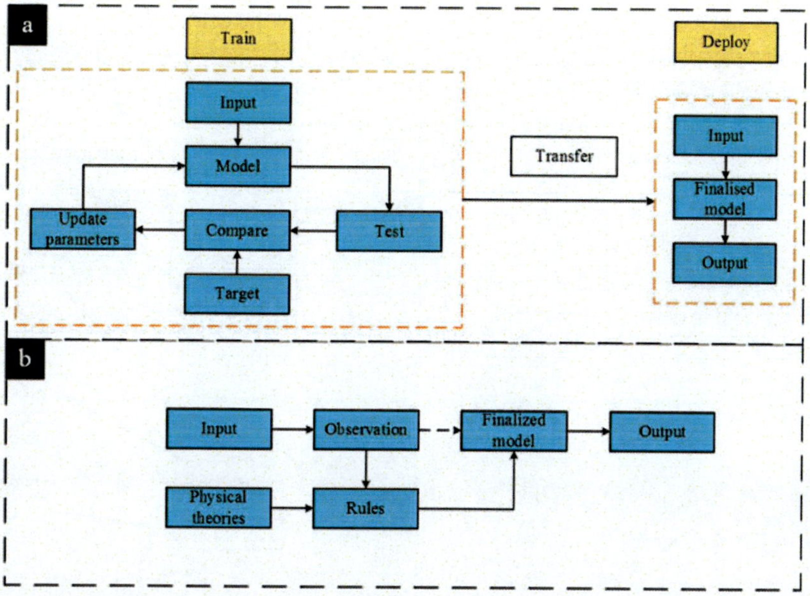

Fig. 2 Modeling method workflow: **a** Machine learning, **b** Traditional

2 Materials and Method

GFRP sheet of 10 mm thickness is used for the drilling operation. The properties of the workpiece material are given in Table 1. A 11 × 11 cm^2 piece was cut out from the sheet and it was clamped on the fixture with the help of two holes. A suitable clearance was given between the workpiece and drilling machine table to avoid damage to the drill bit.

Figure 3 is depicting the schematic of the drilling setup. A conventional radial drilling machine was used to perform the drilling operation of GFRP composite material. To perform this study, the set of experimental runs were planned using the *design of experiment* technique.

Table 1 Properties of W/P material

Properties	Unit	Value
Density	g/cm^3	1.70–1.90
Bending strength	kgf/cm^2	4000–longitudinal 3000–horizontal
Tensile strength	kgf/cm^2	3500–longitudinal 2500–horizontal
Impact strength (Charpy)	kj/m^2	33
Flexural strength	MPa	340

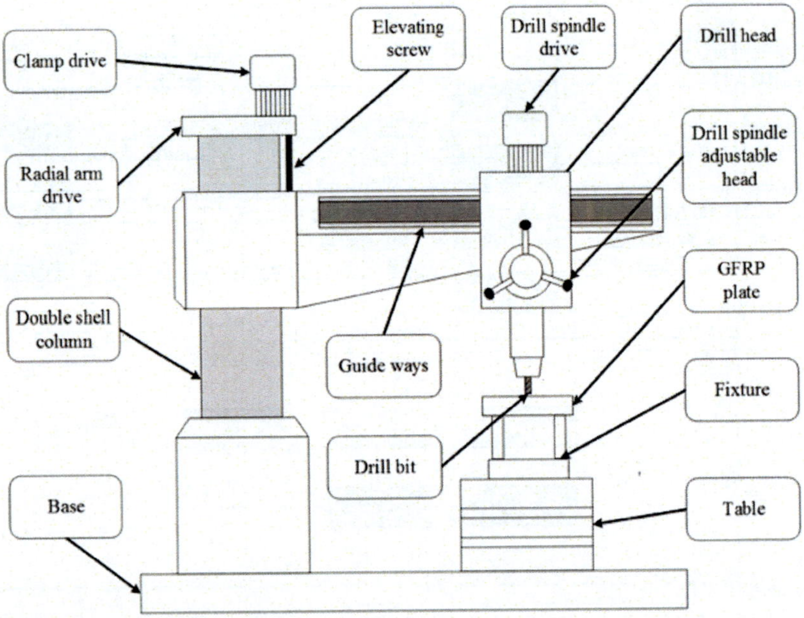

Fig. 3 Schematics of drilling setup

Table 2 Factors and their levels

Factors	Symbols	Units	Levels		
			Level-1	Level-2	Level-3
Feed	A	mm/rev	0.1	0.15	–
Speed	B	rpm	290	580	890
Drill diameter	C	mm	6.0	7.1	9.1
Point angle	D	degree	78	98	118

Taguchi's L_{18} orthogonal array (OA) is implemented for this purpose as it saves resources and helps to draw meaningful conclusions without much loss of information [8]. The different parameters and their levels consider in this study are shown in Table 2.

3 Results and Discussion

The experimental runs were performed as per the L_{18} OA. The outcomes of experimentation are shown in Table 3.

Table 3 Experimental results for MRR

Exp. runs	Feed (mm/rev)	Speed (rpm)	Drill diameter (mm)	Point angle (°)	Material removal rate (cm^3/s)		Residuals
					Experimental	Predicted	
1	0.1	290	6.0	78	0.29	10.7520	0.00
2	0.1	290	7.1	98	0.43	7.3306	0.00
3	0.1	290	9.1	118	0.37	8.6360	0.00
4	0.1	580	6.0	78	0.30	10.4576	0.00
5	0.1	580	7.1	98	0.51	5.8486	0.00
6	0.1	580	9.1	118	0.80	1.9382	−0.01
7	0.1	890	6.0	98	0.38	8.4043	0.00
8	0.1	890	7.1	118	0.25	12.0412	−0.01
9	0.1	890	9.1	78	0.39	8.1787	0.00
10	0.15	290	6.0	118	0.60	4.4370	0.00
11	0.15	290	7.1	78	0.74	2.6154	0.00
12	0.15	290	9.1	98	1.10	−0.8279	0.00
13	0.15	580	6.0	98	0.60	4.4370	0.00
14	0.15	580	7.1	118	0.50	6.0206	−0.04
15	0.15	580	9.1	78	0.75	2.4988	−0.01
16	0.15	890	6.0	118	0.54	5.3521	0.00
17	0.15	890	7.1	78	0.44	7.1309	0.01
18	0.15	890	9.1	98	0.96	0.3546	0.00

Figure 4 is depicting the effect of the machining parameters on MRR. The MRR during drilling of GFRP is increasing with a higher level of speed and drill diameter. The feed and point angle curves have less slop, so they have less effect on MRR. The higher level of feed is directly proportional to MRR, while point angle is having an adverse relationship with MRR. The effect of point angle is complex, and in this investigation, it is observed that the lowest level of feed is best suited for higher MRR.

The analysis of variance (ANOVA) is also performed to quantify the effect of each machining parameters on MRR. The study was conducted at a statistical confidence level of 95%, and the results of ANOVA analysis are summarized in Table 4. As we found the speed followed by drill diameter having the highest impact on MRR, their contribution is found to be 49.30% and 30.99%, respectively. Feed also found significant parameter in this study but is having an 8.52% contribution only. While the contribution of point angle is negligible (1.17%), and the effect of this on MRR can be neglected.

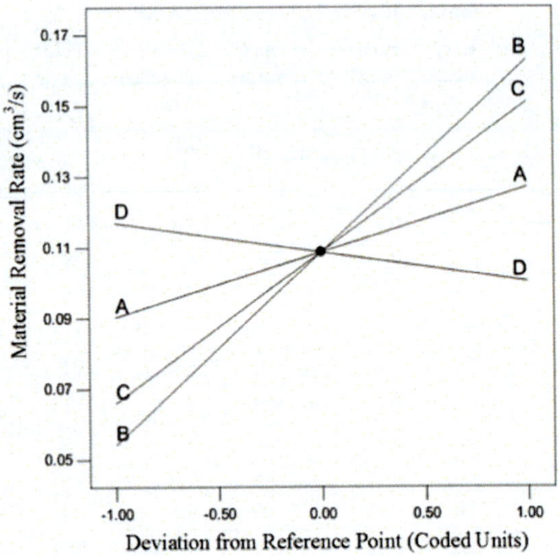

Fig. 4 Effect of machining parameters on MRR

Table 4 ANOVA results for MRR

Source	Sum of squares	df	Mean square	F value	p-value Prob > F	% contribution	Remarks
Model	0.065	4	0.016	30.52	<0.0001		Significant
A	6.050E-003	1	6.050E-003	11.44	0.0049	8.52	Significant
B	0.035	1	0.035	66.98	<0.0001	49.30	Significant
C	0.022	1	0.022	42.06	<0.0001	30.99	Significant
D	8.333E-004	1	8.333E-004	1.58	0.2315	1.17	In significant
Residual	6.875E-003	13	5.289E-004			10.02	
Cor Total	0.071	17					

4 Artificial Neural Network Modeling

Artificial neural network (ANN) mimics the human brain and is a kind of artificial intelligence (AI) techniques. These networks can model both linear and non-linear systems efficiently with no requirement of much mathematical background. ANNs are an information processing system and consist of *neurons* functioning in unison to decipher unambiguous problems. Figure 5 is depicting a simple neural network model that consists of three primary elements; synapses weights, summing node, and an externally applied bias to control the net input of the activation function. The value of weights and biases in every single neuron determines the performance of the network.

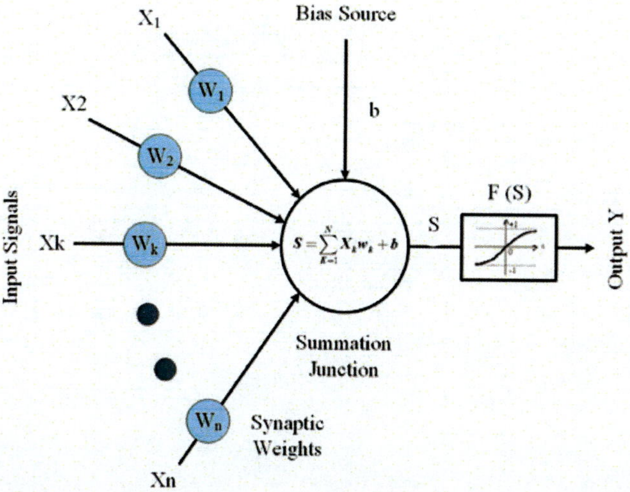

Fig. 5 Simple neuron model

An ANN model in its simplest form consists of an input, output, and hidden layer. But the number of layers and neurons in them can be varying depending upon the problem to be solved. The method of input data required and network design determines the selection of layers for an ANN.

After configuring the ANN, it is trained using the training data set and its performance is evaluated. The training goes so on until the desired outcome from the ANN is not obtained using input data sets. If it is not the case, the weights are adjusted in such a way that the error function minimizes the variation between experimental and predicted values and follows Eq. 1 [3]. This auto-correction is done by the back-propagation (BP) algorithm. The predicted values from the ANN model are again compared with the experimental data. If there is a minute difference between the network's output and experimental output, this process is stopped means errors are under the desired tolerance limit [7].

$$y_k = \varphi\left[\sum_{j=1}^{m} w_{kj} + b_k\right] \tag{1}$$

where k and j denote the neuron and synapse number, respectively.

Figure 6 shows the architecture of the BP neural network used for the problem. The neuron in the input layer receives input. Each input is multiplied by connection weight. Hidden nodes sum the weighted input signals and apply the activation function and send this signal to all the nodes in the layer above, i.e., to the output layer. Each output neuron sums its weighted input signals and applies an activation function to calculate the output signal. Activation functions used for the hidden layer

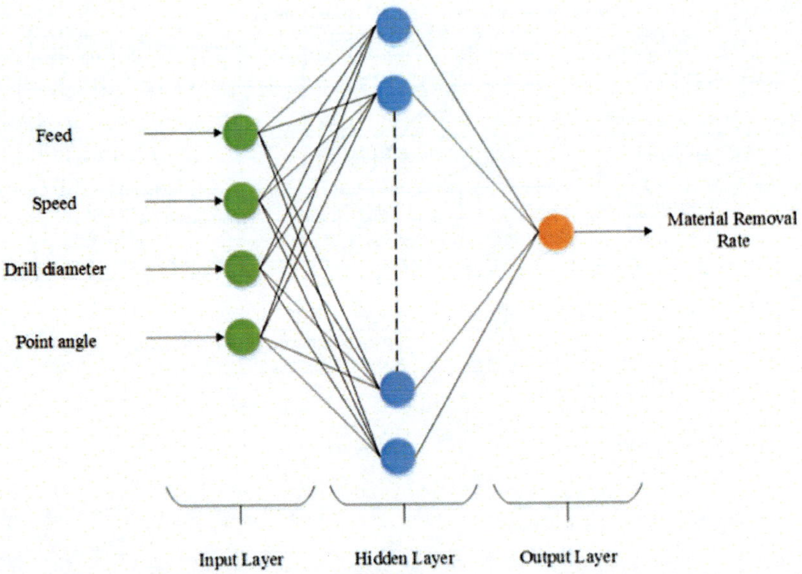

Fig. 6 An artificial neural network architecture (4–8-1)

and output layer are tan sigmoid and linear activation functions, respectively. Each output neuron calculates the difference between target corresponding to input and output and propagates the error back, and modifies the weights. Delta rule is used for updating weights. According to this rule, the adjustment made to the synaptic weights of the neuron is proportional to the product of the error signal and the input signal of the synapse, as shown in Eq. 2.

$$\Delta W_i = \eta(t - Y_i)X_i \tag{2}$$

where η is the learning rate, and y_{in} is the input to the output neurons, t is the target vector.

The input and the corresponding output are presented to the network until it learned the desired relationship. In this research work, feed, speed, drill diameter, and point angle are defined as input parameters, and MRR is taken as an output parameter. The input layer with four nodes and an output layer with one node is considered for the present case. The nodes in the hidden layer are varied from 3 to 10. The optimum numbers of nodes in the hidden layer are found using the trial-and-error approach. New unseen data are exposed to the model to evaluate the performance of the developed model. The training data set used for training the ANN are shown in Table 5. About 70% of the data is used for training, 15% for validation, and 15% data for testing.

Table 5 Training data set

Runs	Feed (mm/rev)	Speed (rpm)	Drill diameter (mm)	Point angle (°)	MRR (cm^3/s)
1	0.1	290	6.0	78	0.02
2	0.1	290	7.1	98	0.04
3	0.1	290	9.1	118	0.06
5	0.1	580	7.1	98	0.05
7	0.1	890	6.0	98	0.08
9	0.1	890	9.1	78	0.12
10	0.15	290	6.0	118	0.08
11	0.15	290	7.1	78	0.13
12	0.15	290	9.1	98	0.19
13	0.15	580	6.0	98	0.04
16	0.15	890	6.0	118	0.08
18	0.15	890	9.1	98	0.09

Figure 7 is depicting the regression plots for the developed neural network outputs with respect to experimentally obtained MRR of drilling operation for training data sets.

After obtaining the configured network, it is tested, and its performance is evaluated by creating a regression plot, which shows the relationship between the outputs of the ANN and the experimental values. For a perfect ANN, the predicted and experimental values would be exactly equal, but such a relationship is rarely perfect in actual practice. Table 6 is showing the data sets used to test the ANN.

Figure 8 is depicting the regression plots for the developed neural network outputs with respect to experimentally obtained MRR of drilling operation for testing data sets.

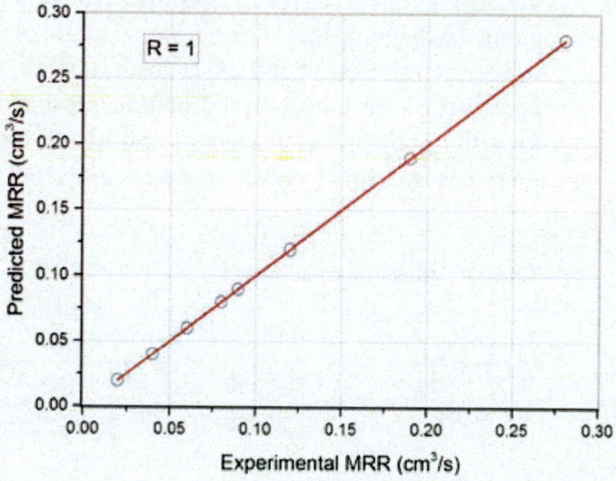

Fig. 7 Regression plot for training data

Table 6 Testing data set

Runs	Feed (mm/rev)	Speed (rpm)	Drill diameter (mm)	Point angle (°)	MRR (cm^3/s)
6	0.1	580	9.1	118	0.80
8	0.1	890	7.1	118	0.25
15	0.15	580	9.1	78	0.75

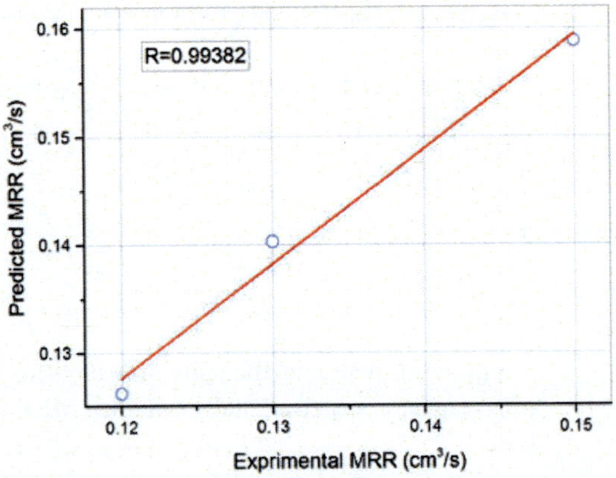

Fig. 8 Regression plot for testing data

After obtaining the configured ANN, it is validated, and its performance is evaluated by creating a regression plot. Table 7 is showing the data sets used to validate the ANN.

Figure 9 is depicting the regression plots for the developed ANN outputs with respect to experimentally obtained MRR of drilling operation for validation data sets.

Figure 10 is depicting the regression plot for the overall ANN performance for the drilling of GFRP using HSS drill bits. It is easily observable that data is aligned along the diagonal line. There is no variation or spread along the diagonal line plotted in between predicted and experimental MRR, and the ANN outcomes are in good agreement with the observed pattern present in the data set. Therefore, the fit is logically good as initial weights and biases for ANN are perfectly arranged for all three cases, with R values in each case close to unity (more than 0.9). If $R = 1$

Table 7 Validation data set

Runs	Feed (mm/rev)	Speed (rpm)	Drill diameter (mm)	Point angle (°)	MRR (cm^3/s)
4	0.1	580	6.0	78	0.30
14	0.15	580	7.1	118	0.50
17	0.15	890	7.1	78	0.44

Fig. 9 Regression plot for validation data

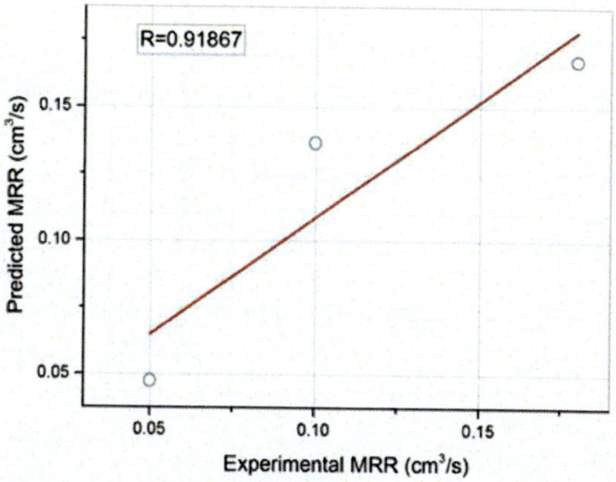

Fig. 10 Regression plot total data

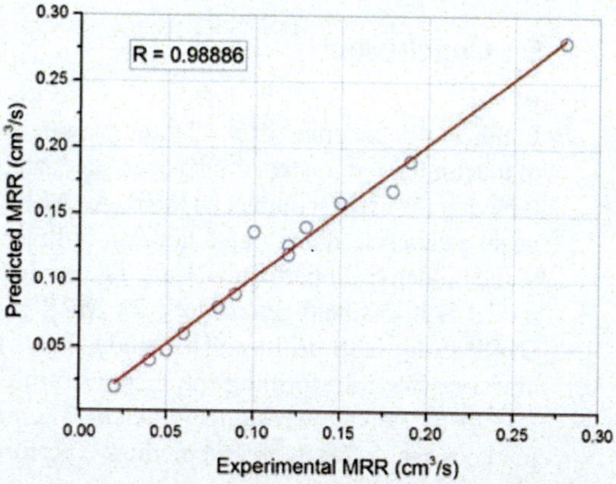

represents an exact linear relationship between ANN predicted and experimentally obtain MRR values. If R is close to zero, then there is no linear relationship between predicted and experimental values.

To verify the goodness of the predicted model, the experimental and ANN predicted MRR are plotted against experimental runs, as shown in Fig. 11. It is easily despicable that the maximum and minimum error variation for MRR is satisfactory. The ANN model fitted well and MAPE for the present study is 3.749 only. This error is permissible for the use of developed ANN model and made it adequate for prediction purposes.

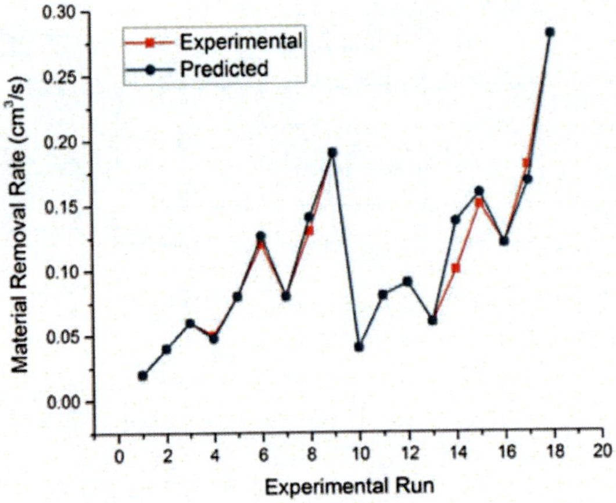

Fig. 11 Experimental versus predicted MRR

5 Conclusions

In this work, the effect of machining parameters on MRR is investigated. The result of machining parameters is also quantified, and it found that speed and drill diameter is having the highest impact on MRR. An AI-based model also has been implemented using a supervised learning algorithm (offline). An ANN model with eight neurons in the hidden layer performed well. The prediction efficiency of the developed ANN model is also found satisfactory for MRR prediction for conventional drilling of GFRP using HSS drill bit. This study shows the ANN capability for Industry 4.0 envisions for transforming the manufacturing sector to make it more productive, less costly, and more resource-efficient. Therefore it is found that machine learning methods may offer improved predictive performance in the GFRP drilling process.

References

1. Akdemir D, Jannink J-L (2014) Ensemble learning with trees and rules: Supervised, semi-supervised, unsupervised. Intel Data Anal 18(5):857–872
2. Bakis CE, Bank LC, Brown V, Cosenza E, Davalos J, Lesko J, Machida A, Rizkalla S, Triantafillou T (2002) Fiber-reinforced polymer composites for construction—state-of-the-art review. J Compos Constr 6(2):73–87
3. Haykin S (2010) Neural networks and learning machines. 3/E, Pearson Education India
4. Hocheng H, Tsao C (2006) Effects of special drill bits on drilling-induced delamination of composite materials. Int J Mach Tools Manuf 46(12–13):1403–1416
5. Ismail SO, Dhakal HN, Popov I, Beaugrand J (2016) Comprehensive study on machinability of sustainable and conventional fibre reinforced polymer composites. Int J Eng Sci Technol 19(4):2043–2052
6. Murphy KP (2012) Machine learning: a probabilistic perspective. MIT Press
7. Rojas R (2013) Neural networks: a systematic introduction. Springer

8. Ross PJ (1988) Taguchi techniques for quality engineering: loss function, orthogonal experiments, parameter and tolerance design. McGraw-Hill, New York.
9. Shalev-Shwartz S, Ben-David S (2014) Understanding machine learning: from theory to algorithms. Cambridge University Press
10. Shunmugesh K, Panneerselvam K (2016) Machinability study of carbon fiber reinforced polymer in the longitudinal and transverse direction and optimization of process parameters using PSO–GSA. Int J Eng Sci Technol 19(3):1552–1563
11. Vankanti VK, Ganta V (2014) Optimization of process parameters in drilling of GFRP composite using Taguchi method. J Market Res 3(1):35–41

Influence of Warm Forging on Mechanical and Microstructural Properties of 316L Steel

Viranshu Kumar, Manohar Kumar Singh, Ghanshyam Das, and Ratnesh Kumar Gupta

1 Introduction

The thermal reactor is one of the most powerful reactors used in thermal plants to produce electricity and thermal energy [1]. High temperatures reactors and pressurized water reactors (HTRs & PWR) are reactors made from 316L used in thermal plants [1, 2]. Microstructure and mechanical property investigation of thermal reactor materials are highly important for a better understanding of plastic deformation through warm deformation [3–5]. At high-temperature service, alloying elements are not stable which is caused by microstructural change at high temperature. Alloying elements' instability at high temperatures also changes microstructure and mechanical properties. Corrosion is also responsible for the degradation of mechanical and microstructural properties, so it needs to be evaluated at high-temperature service conditions [5–8]. Plastic deformation through warm forging changes the microstructure of the material which degraded the mechanical properties of forged steel [9, 10]. It is thus important to control allowing element thermodynamic tendency and microstructural change, which is caused by plastic deformation at high temperatures [10, 11]. Rolling and forging is a very useful way to modify mechanical and microstructural properties of stainless steel because it modifies the microstructure of material by changing the orientation and size of the grain. Among all the plastic deformation processes most cost-effective process is forging which improves the corrosion, microstructural, and mechanical properties of 316L by reducing the size

V. Kumar (✉)
Arka Jain University, Jamshedpur, Jharkhand, India
e-mail: viranshu.k@arkajainuniversity.ac.in

M. K. Singh
Mechanical Engineering Department, ICFAI University, Ranchi 835222, India

G. Das · R. K. Gupta
Materials and Metallurgical Engineering Department, NIFFT, Hatia, Ranchi 834003, India

R. Prasad et al. (eds.), *Advancement in Materials Processing Technology*,
Springer Proceedings in Materials 12,
https://doi.org/10.1007/978-981-16-3297-6_16

Table 1 Composition of AISI 316L steel

Elements	C	P	Si	Mn	Cr	Ni	Mo	N	Fe
Wt %	0.025	0.021	0.494	1.10	16.51	9.80	1.30	0.10	Bal

of grain and changing the orientation of grain [12–15]. Mechanisms of plastic deformation which influence the mechanical and microstructural properties need to be evaluated. In the present investigations, we study the effect of plastic deformation and annealing on the mechanical strength and microstructure of the AISI 316L steel.

2 Experimental Methods

316L steel was used for the present study and its compositions are shown in Table 1. The plate of steel with a dimension of 50 × 50 × 5 mm^3 was plastically deformed to 0–50% by using a hydraulic press of 150 Ton at 750 °C and cooled in air. After the plastic deformation, some warm forged samples were heat-treated (solution annealed) at 1050 °C for 2 h and cooled in water [4]. Microstructure analysis samples were prepared with a dimension of 15 × 15 × 3 mm^3 and polished with 1 µm size using emery paper and diamond past, respectively. For the EBSD microstructure study, electropolishing needs to be done by using a solution of per-caloric acid and methanol at 16 V for 18 s. Vickers hardness test was performed at a load of 200-g force at 5 respective points at the mid of the sample. Charpy impact test was performed according to ASTM E23—18 methods.

3 Results and Discussion

3.1 Microstructural Evolution

EBSD microstructure after deformation is shown in Fig. 1a, b c. Annealing twins with austenitic structure was revealed in the as-received sample while ultra-fine grain size was observed in 50% deformed sample. From Fig. 1b it is observed that grain boundaries are not visible which is due to fine grain formation after deformation. Figure 1c shows the EBSD-SEM microstructure of the heat-treated sample at 1050 °C and cooled in water immediately after 50% deformation. From Fig. 1c we found that heat treatment at 1050 °C causes grain coarsening. After annealing of the sample recrystallization of austenite grains occurs, this leads to the formation of strain-free and equiaxed grains. As per the literature, homogenized grains after heat- treatment gives optimum mechanical strength [4]. EBSD analysis shows that dynamic

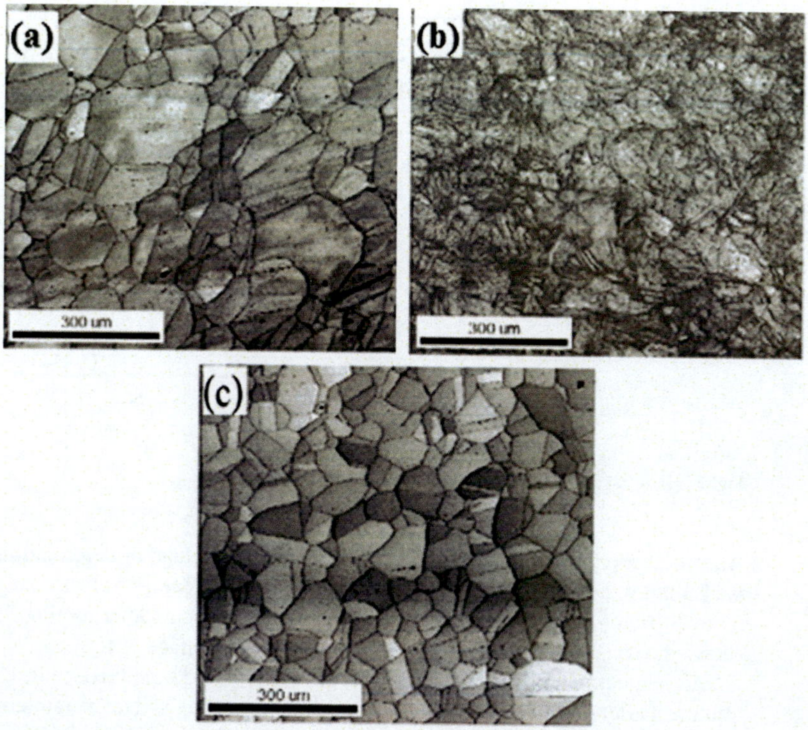

Fig. 1 EBSD-SEM microstructure of **a** as received **b** 50% warm deformed **c** heat-treated at 1050 °C.

recrystallization occurs due to an increase in the fraction of low-angle grain boundaries after deformation. The mechanical property of the material depends upon the microstructure of the material which is developed during service conditions.

3.2 Effect of Warm Deformation on Hardness and Impact Strength

As per the literature survey, plastic deformation changes the hardness and toughness of the material, and to obtain the optimum toughness and hardness heat treatment is done after deformation [4]. Five repetitive tests were performed for the Impact test and Vickers hardness test, respectively, and the average value was taken for optimum value for impact strength and hardness, respectively. From the Vickers hardness test, we observed that hardness values change from 180 to 305 VHN which is not good for an industrial application so it needs to revel with the help of heat treatment. After heat treatment at 1050 °C of 50% deformed sample, hardness becomes 245 VHN.

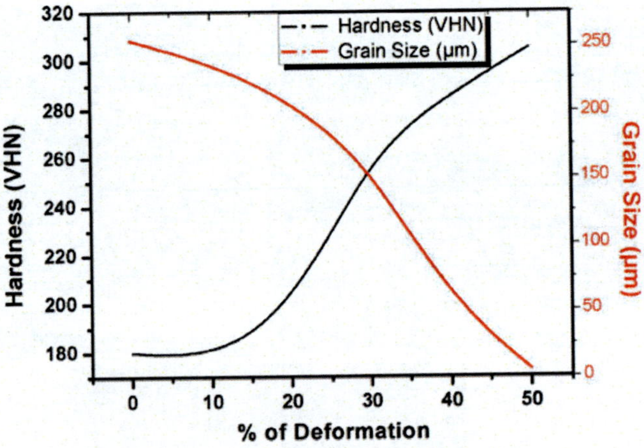

Fig. 2 Hardness and grain size

Absorbed energy decreases with increases in the percentage of deformation which is also not good for automobile structure, and absorbed energy values change from 55 to 30 J/cm^2 which indicates that loss of toughness occurs after forging. Figure 2 shows the changes in grain size hardness with the percentage of forging.

Mechanical and microstructural properties of deformed samples depend upon heat treatment after deformation because appropriate annealing and temperature improve the mechanical and microstructural properties. The value of residual stress is very high in a 50% deformed sample which is not good for engineering applications so it needs to revel after forging. To obtain the optimum properties of the material, residual stress and strain need to be homogenously distributed after deformation. In this investigation, we found that minimum residual stress and homogeneous distribution of strain were obtained in the annealed sample after deformation. From EBSD analysis we found that grain refinement occurs after forging which increases the strength of the material but same time the value of stresses, strains, and dislocation density increases which are not favourable for the engineering component. In this study, we use heat treatment for relieving the strain and stress in the forged sample. Figure 3 shows the Impact strength and Vickers hardness of the deformed sample.

3.3 *Effect of Deformation on Grain Size*

According to the deformation theory, if the deformation percentage increases the size of the grain decrease which improves the strength of the material. Small grain size materials have high strength but low ductility which is not favourable for an industrial application so annealing is performed after deformation. Solution annealing after

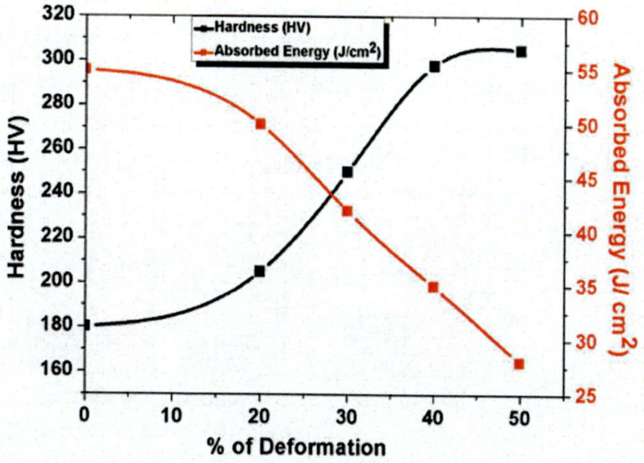

Fig. 3 Variation of hardness and absorbed energy

deformation cause coarsening of grain which gives optimum impact strength and ductility. Figure 4 shows the variation of grain size.

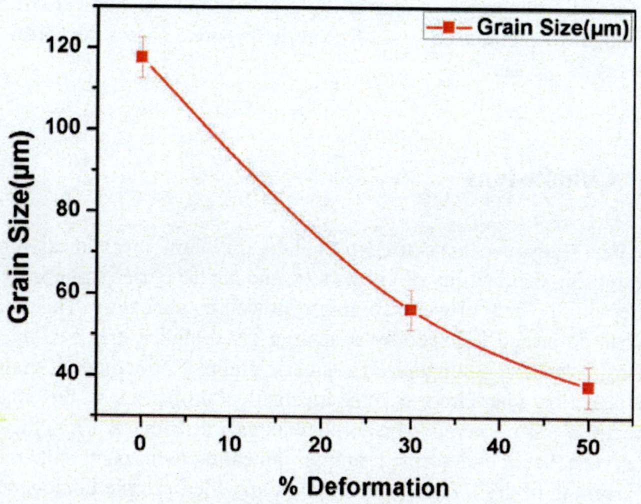

Fig. 4 Grain size of deformed 316L steel

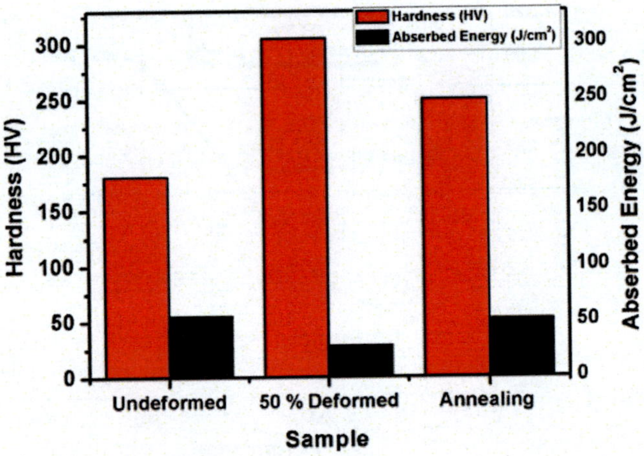

Fig. 5 Hardness and impact strength of deformed 316L steel

3.4 Effect of Forging and Annealing on Mechanical Strength

Solution annealing gives the optimum hardness and impact strength because after deformation appropriate annealing improves the mechanical strength by forming homogeneous equiaxed strain-free grain. Figure 5 shows the hardness and absorbed energy of the sample.

4 Conclusions

In this study, we have tried to obtain optimum mechanical properties without decreasing the ductility of given steel, and we have tried to improve the microstructure with the help of warm forging followed by annealing. The obtained results are Warm deformed followed by annealing gives higher strength due to the formation of equiaxed and homogenized austenite grains. Refinement of grains occurs due to deformation which improves the mechanical properties. In the heat-treated sample, the strain was observed to be homogeneously distributed. From the EBSD result, we observed that in undeformed samples annealing twins occur but after heat-treatment coarsening of grain occurs. In solution annealed sample homogeneous distribution of grain was observed, which produces optimum ductility and impact strength. The homogeneous distribution of grains in the solution annealed samples also improves the hardness value.

References

1. McGuire MF (2008) Stainless steels for design engineers. ASM International
2. Zinkle SJ, Was GS (2013) Materials challenges in nuclear energy. Acta Mater 6:735–758
3. Allen T, Busby J, Meyer M, Petti D (2010) Materials challenges for nuclear systems. Mater Today 13:14–23
4. YuelingGuo E-H, Wang J (2010) Effects of forging and heat treatments on the microstructure and oxidation behavior of 316LN stainless steel in high-temperature water. J Mater Sci Technol 31:403–412
5. Ghosh SK, ShikharJha P, Mallick PP, Chattopadhyay (2013) Influence of mechanical deformation and annealing on kinetics of martensite in a stainless steel. Mater Manuf Proc 28:249–255
6. Chen J, Lu Z, Xiao Q, Ru X, Han G, Chen Z, Zhou B, Shoji T (2016) The effects of cold rolling orientation and water chemistry on stress corrosion cracking behavior of 316L stainless steel in simulated PWR water environments. J Nuclear Mater 472
7. Chio JY, Jin W (1997) Strain-induced martensite formation and its effect on strain hardening behavior in the cold drawn 304 austenitic stainless steels. Scripta Mater 36:99–104
8. Zhoua Y, Liua Y, Zhoua X (2017) Precipitation, and hot deformation behavior of austenitic heat-resistant steels: a review. J Mater Sci Technol
9. Olson GB, Cohen M (1972) A mechanism for the strain-induced nucleation of martensitic transformations. J Less Common Metals 28:107–118
10. Maxwell PC, Goldberg A, Shyne JC (1974) Stress-assisted and strain-induced martensite in Fe-Ni-C alloys. Metallurgic Trans 5:1305–1318
11. Lee WS, Lin CF (2000) The morphologies and characteristics of impact induced martensite in 304L stainless steel. Scripta Materialia 43:777–782
12. Spencer K, Embury JD, Konlon KD, Veron M, Brechet Y (2004) Strengthening via the formation of strain-induced martensite in stainless steel. Mater Sci Eng A 387–389
13. Kumara V, Gupta RK, Das G (2020) Influence of deformation and annealing on microstructure and corrosion behavior of austenitic stainless steel. Mater Today Proc 22(2020):3347–3352. https://doi.org/10.1016/j.matpr.2020.03.361
14. Kumar V, Gupta R, Das G (2020) Effect of warm forging on the microstructure and corrosion behavior of austenitic stainless steel 316LN. Mater Perform Charact 9(1):94–106. https://doi.org/10.1520/MPC20190228
15. Kumar V, Gupta RK, Das G (2020) Influence of forging and annealing on the microstructure and corrosion behavior of austenitic stainless steel. J Inst Eng India Ser D 101:105–109. https://doi.org/10.1007/s40033-020-00209-2

Recycling and Reuse of Iron Ore Pellet Fines

P. Prusti, K. Barik, D. K. Sahu, S. Soren, B. C. Meikap, and S. K. Biswal

1 Introduction

Iron ores are valuable natural resources being finite and non-renewable. Iron ore is one of the basic raw materials for iron and steel industries. It is predicted that the world steel production may reach more than 2600 million tonnes and in India it may reach 300 million tonnes by 2030 [1]. This has resulted in a large increase in the demand for iron ore. It is a great concern for steel industries to either receive suitable quality iron ore in the form of lumpy or sinter/pellet. To utilize the ultra-fine particles after beneficiation of low and lean grade resources, pelletization is the only alternative technology. Therefore, a greater number of pelletizing plants have been set up all over the world having capacity more than 800 million tonnes per annum to utilize iron ore high-grade fines to improve the conservation of iron ore in the world.

During pelletization process and preparation of feed for DRI or reduction smelting processes, around 3% of pellet fines (less than 8 mm) is generated as undesired product in existing pelletization plants. The quantity will be more than 24 million

P. Prusti (✉) · D. K. Sahu · S. K. Biswal
Mineral Processing Department, CSIR-Institute of Minerals and Materials Technology, Bhubaneswar 751013, India
e-mail: pprusti@immt.res.in

P. Prusti · B. C. Meikap
Department of Chemical Engineering, Indian Institute of Technology Kharagpur, Kharagpur 721302, India

K. Barik
Department of Chemical Engineering, Indira Gandhi Institute of Technology Sarang, Dhenkanal 759146, India

K. Barik · S. Soren
Department of Fuel and Minerals, Metallurgical Engineering, Indian Institute of Technology (ISM) Dhanbad, Dhanbad 826004, India

R. Prasad et al. (eds.), *Advancement in Materials Processing Technology*, Springer Proceedings in Materials 12,
https://doi.org/10.1007/978-981-16-3297-6_17

References

1. McGuire MF (2008) Stainless steels for design engineers. ASM International
2. Zinkle SJ, Was GS (2013) Materials challenges in nuclear energy. Acta Mater 6:735–758
3. Allen T, Busby J, Meyer M, Petti D (2010) Materials challenges for nuclear systems. Mater Today 13:14–23
4. YuelingGuo E-H, Wang J (2010) Effects of forging and heat treatments on the microstructure and oxidation behavior of 316LN stainless steel in high-temperature water. J Mater Sci Technol 31:403–412
5. Ghosh SK, ShikharJha P, Mallick PP, Chattopadhyay (2013) Influence of mechanical deformation and annealing on kinetics of martensite in a stainless steel. Mater Manuf Proc 28:249–255
6. Chen J, Lu Z, Xiao Q, Ru X, Han G, Chen Z, Zhou B, Shoji T (2016) The effects of cold rolling orientation and water chemistry on stress corrosion cracking behavior of 316L stainless steel in simulated PWR water environments. J Nuclear Mater 472
7. Chio JY, Jin W (1997) Strain-induced martensite formation and its effect on strain hardening behavior in the cold drawn 304 austenitic stainless steels. Scripta Mater 36:99–104
8. Zhoua Y, Liua Y, Zhoua X (2017) Precipitation, and hot deformation behavior of austenitic heat-resistant steels: a review. J Mater Sci Technol
9. Olson GB, Cohen M (1972) A mechanism for the strain-induced nucleation of martensitic transformations. J Less Common Metals 28:107–118
10. Maxwell PC, Goldberg A, Shyne JC (1974) Stress-assisted and strain-induced martensite in Fe-Ni-C alloys. Metallurgic Trans 5:1305–1318
11. Lee WS, Lin CF (2000) The morphologies and characteristics of impact induced martensite in 304L stainless steel. Scripta Materialia 43:777–782
12. Spencer K, Embury JD, Konlon KD, Veron M, Brechet Y (2004) Strengthening via the formation of strain-induced martensite in stainless steel. Mater Sci Eng A 387–389
13. Kumara V, Gupta RK, Das G (2020) Influence of deformation and annealing on microstructure and corrosion behavior of austenitic stainless steel. Mater Today Proc 22(2020):3347–3352. https://doi.org/10.1016/j.matpr.2020.03.361
14. Kumar V, Gupta R, Das G (2020) Effect of warm forging on the microstructure and corrosion behavior of austenitic stainless steel 316LN. Mater Perform Charact 9(1):94–106. https://doi.org/10.1520/MPC20190228
15. Kumar V, Gupta RK, Das G (2020) Influence of forging and annealing on the microstructure and corrosion behavior of austenitic stainless steel. J Inst Eng India Ser D 101:105–109. https://doi.org/10.1007/s40033-020-00209-2

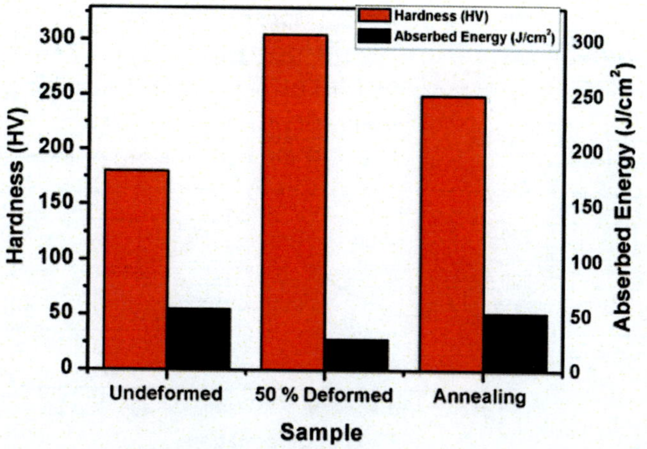

Fig. 5 Hardness and impact strength of deformed 316L steel

3.4 Effect of Forging and Annealing on Mechanical Strength

Solution annealing gives the optimum hardness and impact strength because after deformation appropriate annealing improves the mechanical strength by forming homogeneous equiaxed strain-free grain. Figure 5 shows the hardness and absorbed energy of the sample.

4 Conclusions

In this study, we have tried to obtain optimum mechanical properties without decreasing the ductility of given steel, and we have tried to improve the microstructure with the help of warm forging followed by annealing. The obtained results are Warm deformed followed by annealing gives higher strength due to the formation of equiaxed and homogenized austenite grains. Refinement of grains occurs due to deformation which improves the mechanical properties. In the heat-treated sample, the strain was observed to be homogeneously distributed. From the EBSD result, we observed that in undeformed samples annealing twins occur but after heat-treatment coarsening of grain occurs. In solution annealed sample homogeneous distribution of grain was observed, which produces optimum ductility and impact strength. The homogeneous distribution of grains in the solution annealed samples also improves the hardness value.

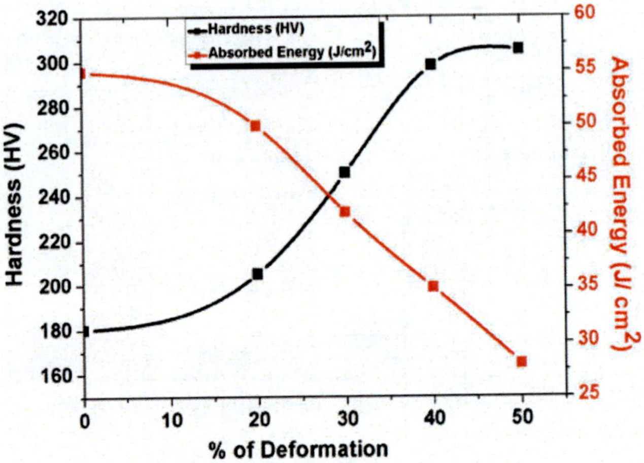

Fig. 3 Variation of hardness and absorbed energy

deformation cause coarsening of grain which gives optimum impact strength and ductility. Figure 4 shows the variation of grain size.

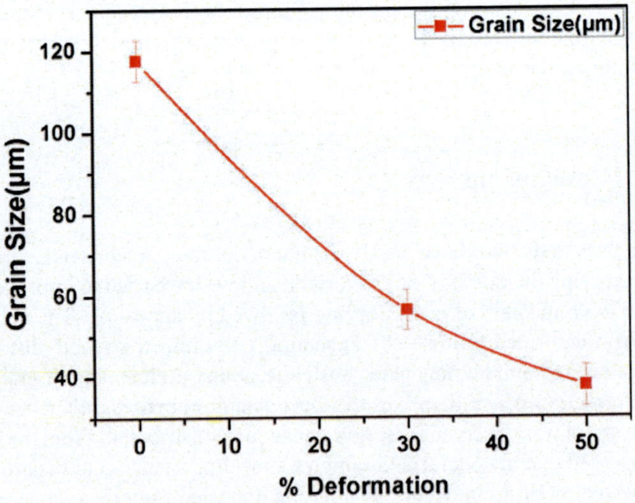

Fig. 4 Grain size of deformed 316L steel

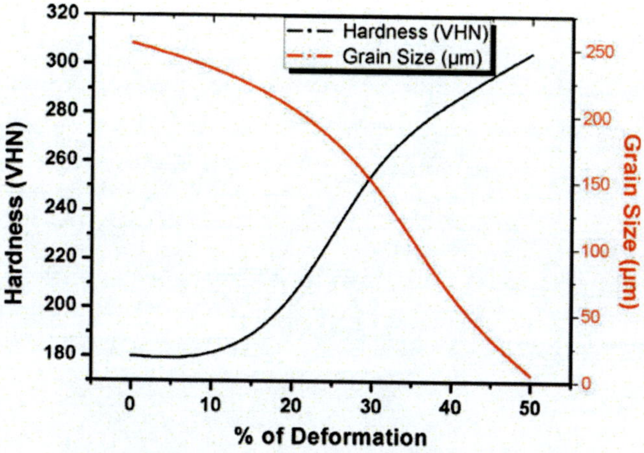

Fig. 2 Hardness and grain size

Absorbed energy decreases with increases in the percentage of deformation which is also not good for automobile structure, and absorbed energy values change from 55 to 30 J/cm^2 which indicates that loss of toughness occurs after forging. Figure 2 shows the changes in grain size hardness with the percentage of forging.

Mechanical and microstructural properties of deformed samples depend upon heat treatment after deformation because appropriate annealing and temperature improve the mechanical and microstructural properties. The value of residual stress is very high in a 50% deformed sample which is not good for engineering applications so it needs to revel after forging. To obtain the optimum properties of the material, residual stress and strain need to be homogenously distributed after deformation. In this investigation, we found that minimum residual stress and homogeneous distribution of strain were obtained in the annealed sample after deformation. From EBSD analysis we found that grain refinement occurs after forging which increases the strength of the material but same time the value of stresses, strains, and dislocation density increases which are not favourable for the engineering component. In this study, we use heat treatment for relieving the strain and stress in the forged sample. Figure 3 shows the Impact strength and Vickers hardness of the deformed sample.

3.3 *Effect of Deformation on Grain Size*

According to the deformation theory, if the deformation percentage increases the size of the grain decrease which improves the strength of the material. Small grain size materials have high strength but low ductility which is not favourable for an industrial application so annealing is performed after deformation. Solution annealing after

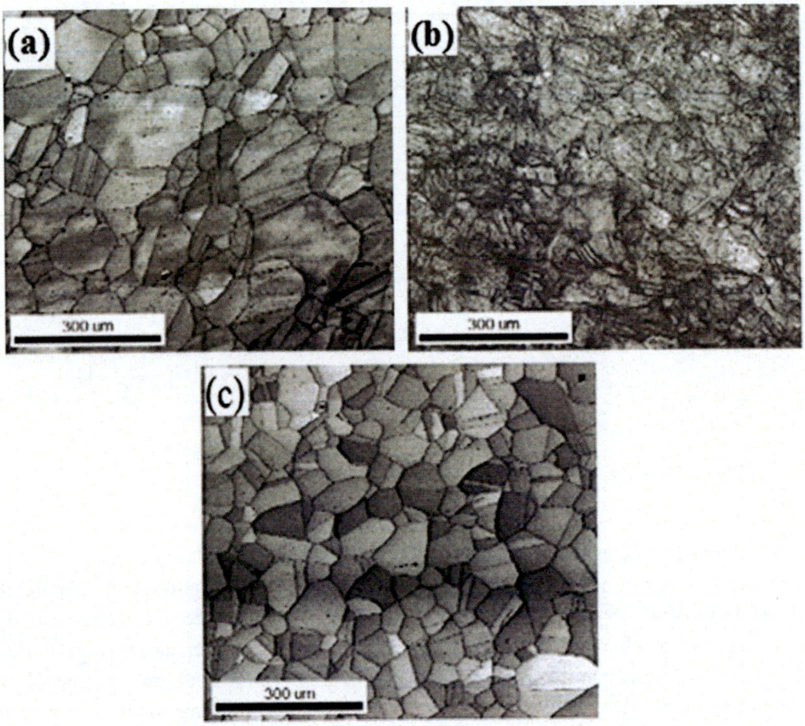

Fig. 1 EBSD-SEM microstructure of **a** as received **b** 50% warm deformed **c** heat-treated at 1050 °C.

recrystallization occurs due to an increase in the fraction of low-angle grain boundaries after deformation. The mechanical property of the material depends upon the microstructure of the material which is developed during service conditions.

3.2 *Effect of Warm Deformation on Hardness and Impact Strength*

As per the literature survey, plastic deformation changes the hardness and toughness of the material, and to obtain the optimum toughness and hardness heat treatment is done after deformation [4]. Five repetitive tests were performed for the Impact test and Vickers hardness test, respectively, and the average value was taken for optimum value for impact strength and hardness, respectively. From the Vickers hardness test, we observed that hardness values change from 180 to 305 VHN which is not good for an industrial application so it needs to revel with the help of heat treatment. After heat treatment at 1050 °C of 50% deformed sample, hardness becomes 245 VHN.

Table 1 Composition of AISI 316L steel

Elements	C	P	Si	Mn	Cr	Ni	Mo	N	Fe
Wt %	0.025	0.021	0.494	1.10	16.51	9.80	1.30	0.10	Bal

of grain and changing the orientation of grain [12–15]. Mechanisms of plastic deformation which influence the mechanical and microstructural properties need to be evaluated. In the present investigations, we study the effect of plastic deformation and annealing on the mechanical strength and microstructure of the AISI 316L steel.

2 Experimental Methods

316L steel was used for the present study and its compositions are shown in Table 1. The plate of steel with a dimension of $50 \times 50 \times 5\ mm^3$ was plastically deformed to 0–50% by using a hydraulic press of 150 Ton at 750 °C and cooled in air. After the plastic deformation, some warm forged samples were heat-treated (solution annealed) at 1050 °C for 2 h and cooled in water [4]. Microstructure analysis samples were prepared with a dimension of $15 \times 15 \times 3\ mm^3$ and polished with 1 μm size using emery paper and diamond past, respectively. For the EBSD microstructure study, electropolishing needs to be done by using a solution of per-caloric acid and methanol at 16 V for 18 s. Vickers hardness test was performed at a load of 200-g force at 5 respective points at the mid of the sample. Charpy impact test was performed according to ASTM E23—18 methods.

3 Results and Discussion

3.1 Microstructural Evolution

EBSD microstructure after deformation is shown in Fig. 1a, b c. Annealing twins with austenitic structure was revealed in the as-received sample while ultra-fine grain size was observed in 50% deformed sample. From Fig. 1b it is observed that grain boundaries are not visible which is due to fine grain formation after deformation. Figure 1c shows the EBSD-SEM microstructure of the heat-treated sample at 1050 °C and cooled in water immediately after 50% deformation. From Fig. 1c we found that heat treatment at 1050 °C causes grain coarsening. After annealing of the sample recrystallization of austenite grains occurs, this leads to the formation of strain-free and equiaxed grains. As per the literature, homogenized grains after heat- treatment gives optimum mechanical strength [4]. EBSD analysis shows that dynamic

Influence of Warm Forging on Mechanical and Microstructural Properties of 316L Steel

Viranshu Kumar, Manohar Kumar Singh, Ghanshyam Das, and Ratnesh Kumar Gupta

1 Introduction

The thermal reactor is one of the most powerful reactors used in thermal plants to produce electricity and thermal energy [1]. High temperatures reactors and pressurized water reactors (HTRs & PWR) are reactors made from 316L used in thermal plants [1, 2]. Microstructure and mechanical property investigation of thermal reactor materials are highly important for a better understanding of plastic deformation through warm deformation [3–5]. At high-temperature service, alloying elements are not stable which is caused by microstructural change at high temperature. Alloying elements' instability at high temperatures also changes microstructure and mechanical properties. Corrosion is also responsible for the degradation of mechanical and microstructural properties, so it needs to be evaluated at high-temperature service conditions [5–8]. Plastic deformation through warm forging changes the microstructure of the material which degraded the mechanical properties of forged steel [9, 10]. It is thus important to control allowing element thermodynamic tendency and microstructural change, which is caused by plastic deformation at high temperatures [10, 11]. Rolling and forging is a very useful way to modify mechanical and microstructural properties of stainless steel because it modifies the microstructure of material by changing the orientation and size of the grain. Among all the plastic deformation processes most cost-effective process is forging which improves the corrosion, microstructural, and mechanical properties of 316L by reducing the size

V. Kumar (✉)
Arka Jain University, Jamshedpur, Jharkhand, India
e-mail: viranshu.k@arkajainuniversity.ac.in

M. K. Singh
Mechanical Engineering Department, ICFAI University, Ranchi 835222, India

G. Das · R. K. Gupta
Materials and Metallurgical Engineering Department, NIFFT, Hatia, Ranchi 834003, India

R. Prasad et al. (eds.), *Advancement in Materials Processing Technology*,
Springer Proceedings in Materials 12,
https://doi.org/10.1007/978-981-16-3297-6_16

8. Ross PJ (1988) Taguchi techniques for quality engineering: loss function, orthogonal experiments, parameter and tolerance design. McGraw-Hill, New York.
9. Shalev-Shwartz S, Ben-David S (2014) Understanding machine learning: from theory to algorithms. Cambridge University Press
10. Shunmugesh K, Panneerselvam K (2016) Machinability study of carbon fiber reinforced polymer in the longitudinal and transverse direction and optimization of process parameters using PSO–GSA. Int J Eng Sci Technol 19(3):1552–1563
11. Vankanti VK, Ganta V (2014) Optimization of process parameters in drilling of GFRP composite using Taguchi method. J Market Res 3(1):35–41

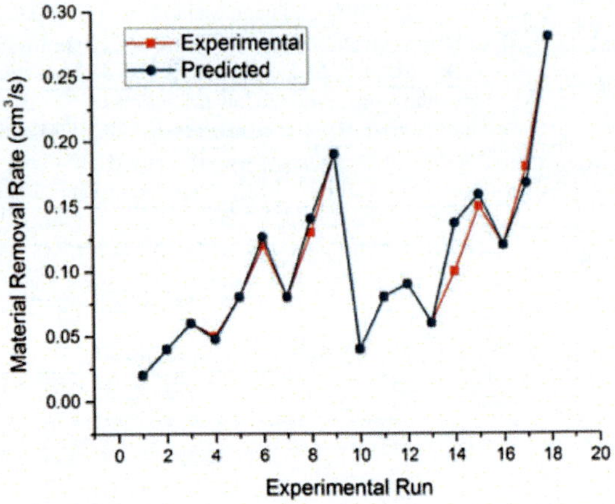

Fig. 11 Experimental versus predicted MRR

5 Conclusions

In this work, the effect of machining parameters on MRR is investigated. The result of machining parameters is also quantified, and it found that speed and drill diameter is having the highest impact on MRR. An AI-based model also has been implemented using a supervised learning algorithm (offline). An ANN model with eight neurons in the hidden layer performed well. The prediction efficiency of the developed ANN model is also found satisfactory for MRR prediction for conventional drilling of GFRP using HSS drill bit. This study shows the ANN capability for Industry 4.0 envisions for transforming the manufacturing sector to make it more productive, less costly, and more resource-efficient. Therefore it is found that machine learning methods may offer improved predictive performance in the GFRP drilling process.

References

1. Akdemir D, Jannink J-L (2014) Ensemble learning with trees and rules: Supervised, semi-supervised, unsupervised. Intel Data Anal 18(5):857–872
2. Bakis CE, Bank LC, Brown V, Cosenza E, Davalos J, Lesko J, Machida A, Rizkalla S, Triantafillou T (2002) Fiber-reinforced polymer composites for construction—state-of-the-art review. J Compos Constr 6(2):73–87
3. Haykin S (2010) Neural networks and learning machines. 3/E, Pearson Education India
4. Hocheng H, Tsao C (2006) Effects of special drill bits on drilling-induced delamination of composite materials. Int J Mach Tools Manuf 46(12–13):1403–1416
5. Ismail SO, Dhakal HN, Popov I, Beaugrand J (2016) Comprehensive study on machinability of sustainable and conventional fibre reinforced polymer composites. Int J Eng Sci Technol 19(4):2043–2052
6. Murphy KP (2012) Machine learning: a probabilistic perspective. MIT Press
7. Rojas R (2013) Neural networks: a systematic introduction. Springer

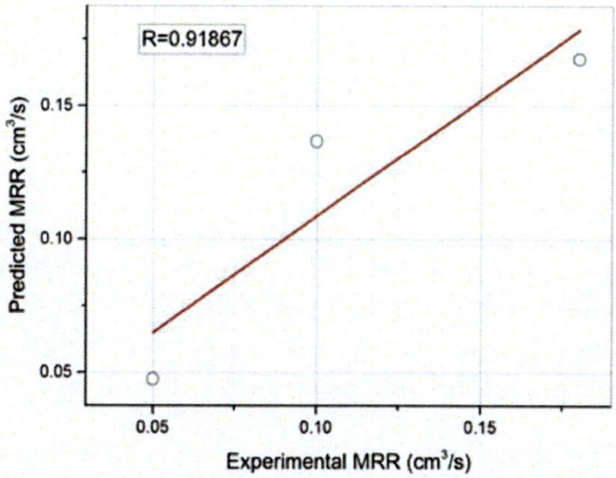

Fig. 9 Regression plot for validation data

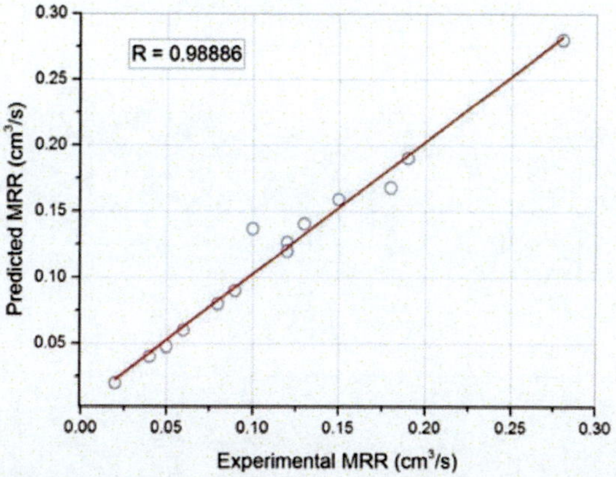

Fig. 10 Regression plot total data

represents an exact linear relationship between ANN predicted and experimentally obtain MRR values. If R is close to zero, then there is no linear relationship between predicted and experimental values.

To verify the goodness of the predicted model, the experimental and ANN predicted MRR are plotted against experimental runs, as shown in Fig. 11. It is easily despicable that the maximum and minimum error variation for MRR is satisfactory. The ANN model fitted well and MAPE for the present study is 3.749 only. This error is permissible for the use of developed ANN model and made it adequate for prediction purposes.

Design, Analysis and Fabrication of Solar-Powered Electric Vehicle for Handicapped—A Sustainable Approach

Manikantan R. Nair, A. Anshadh, R. K. Murali Mohan, S. A. Anu, V. M. Arad, and Mohammed Afsal

1 Introduction

Accessibility is a vital factor to facilitate comfortable lifestyle in present world scenario. According to United Nations, the development should happen in such a way that people with disabilities should also be able to access all the facilities in the nation irrespective of their disabilities [1]. Often transportation means are not much friendly to disabled and thus restrict them from better lifestyle [2]. It was proposed that the environment should be designed, made and continuously modified in such a manner that it always remains friendly to those people with disabilities so that they can also enjoy all the freedom [3]. The significant barriers to the full participation of people with disabilities in the social ecosystem are the inaccessibility which is built into physical environment [4]. Though this was stated long back that, the current environment design was not enough to be friendly towards disabled [5], still the problems are not fully resolved. It is also found that the disabled are not treated as equals in the society [6].

Researchers conclude that travelling and well-being are correlated such that experiences obtained from travelling, various activities carried out during the travel period are helpful in leading a pleasant social life [7, 8]. According to researchers, travelling could provide outcomes like ability to obtain goods and services, access to better healthcare facilities, financial stability, and interactions with various groups of people. Lack of effective transportation system would deny all this to disabled [9]. Studies propose that disabled are often an excluded community in the society [10] and as these people cannot make good earning they are often denied essential life quality [11].

In India, the contemporarily modified scooters used by physically disabled people for transportation are not ergonomic and do not provide protection from adverse

M. R. Nair (✉) · A. Anshadh · R. K. M. Mohan · S. A. Anu · V. M. Arad · M. Afsal
Department of Mechanical Engineering, ACE College of Engineering, Thiruvananthapuram, India

R. Prasad et al. (eds.), *Advancement in Materials Processing Technology*,
Springer Proceedings in Materials 12,
https://doi.org/10.1007/978-981-16-3297-6_18

weather conditions which can be more dangerous for physically disabled people compared to the normal person while driving. This project focuses on providing a feasible design solution in form of a user-friendly three-wheel vehicle, which allows physically challenged or debilitated person with partial disability to commute themselves and perform their activities without anyone's assistance. Recognizing this we were able to make a model of solar-powered electric three-wheeler for disabled people. This model was developed at low cost using sustainable approach. No Emission is also an advantage for this vehicle.

1.1 Problem Definition

The disabled have fewer opportunities and lower life quality due to poor accessibility.

1.2 Objective

To design and fabricate a vehicle for handicapped people such that it would help to solve their transportation problem.

2 Design

CREO parametric 3D modelling tool was used to design the required vehicle model. The dimensions of the vehicle design were calculated with respect to the standard wheelchair dimensions such that a person using wheelchair would be able to board himself into the vehicle along with the wheelchair without any assistance from others and use the vehicle by himself. An automatic ramp mechanism was provided to the side of the vehicle so that the wheelchair can be boarded into the vehicle. The vehicle was designed in such a way that all the controllers were connected to the driving handle in a manner such that the disabled can operate all the functionality using his hands only. A mechanism was designed to lock the wheelchair to the driving area of the vehicle such that the wheelchair may not roll out during the motion of the vehicle (Figs. 1 and 2).

3 Analysis

ANSYS R15.0 was used to perform the analysis. For the process of analysis the mild steel material data, designed model, supports, loads were given as input to the software. The software conducted the analysis upon the model based on the inputs

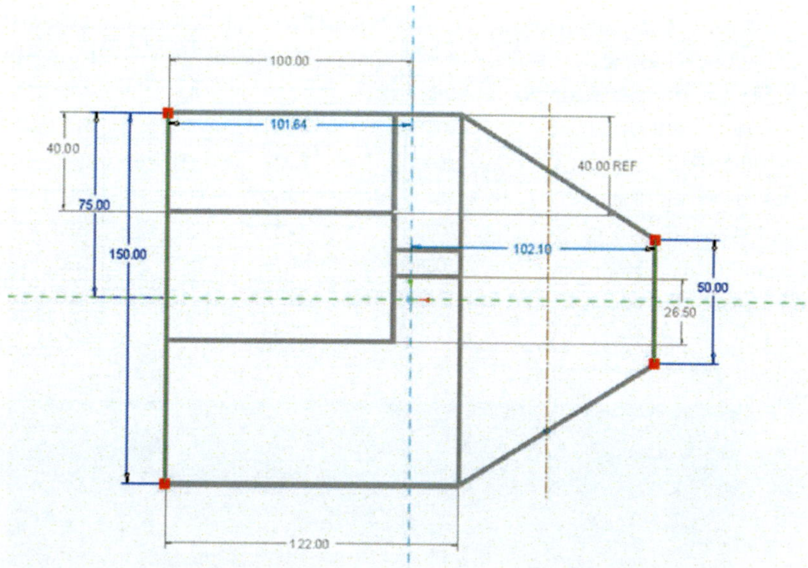

Fig. 1 Design of the base frame

Fig. 2 Vehicle design

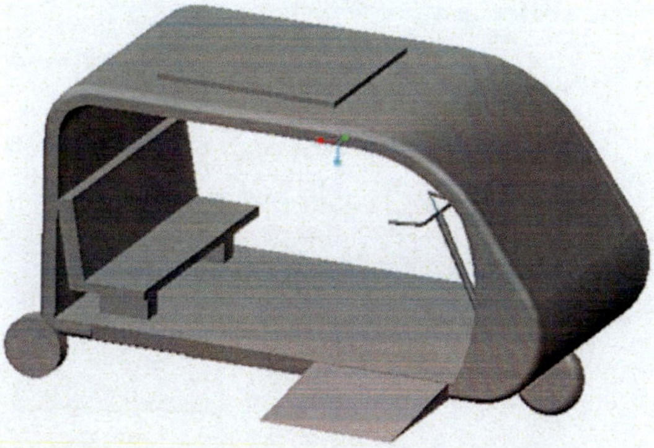

given and gave the resulting deformation and stress values based on the analysis. Two kinds of loads were given as the input to the software, first is the self-weight and latter was the variable load which is the weight of the person using the vehicle. For mild steel the yield stress is 247 MPa. This design is made for a single person with disability, where the stress value as per the analysis is 24.239 MPa only, hence the design is safe (Table 1 and Figs. 3, 4).

Table 1 Comparative analysis for Deformation and Stress with variations in variable load

Sl.no	Load		Deformation (mm)	Stress (MPa)
	Self weight (N)	Others (N)		
1	800	800	6.66	24.239
2	800	1000	7.927	28.239
3	800	3000	20.569	79.478
4	800	6000	39.533	158.5
5	800	10,000	60	263
6	800	11,000	71.139	290.23
7	800	12,000	77.46	316.37

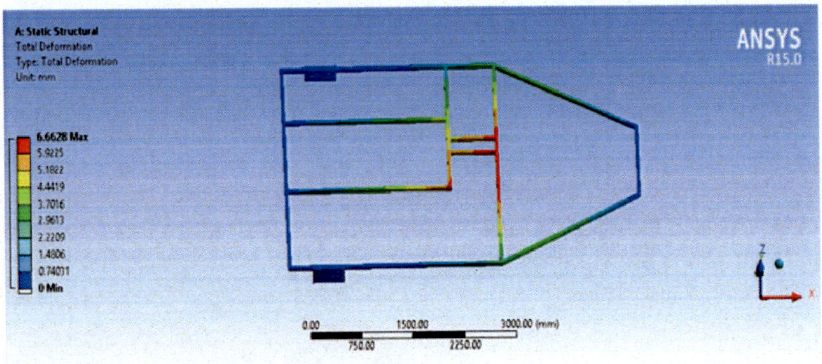

Fig. 3 Deformation at self weight-800 N and others-800 N

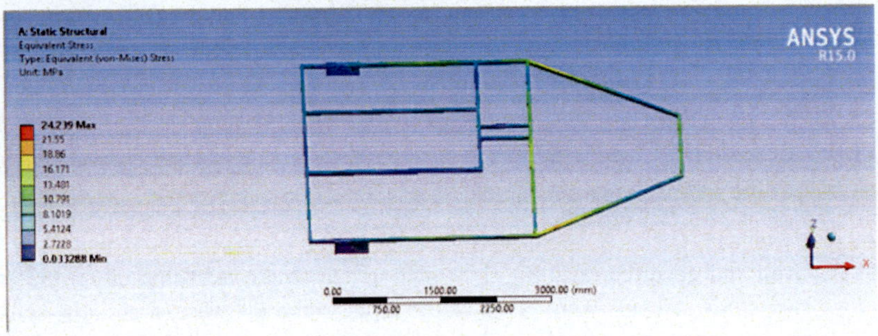

Fig. 4 Stress distribution at self weight-800 N and others-800 N

Table 2 Cost of various materials used in manufacturing process

Sl. no	Materials	Cost (in rupees)
1	Hollow section	600
2	Angle	1000
3	Rectangular bar	1540
4	Sheet metal	300
5	Motor	15,000
6	Battery	4000
7	Glass	1000
8	Solar panel	6000

4 Cost Analysis

The cost analysis was conducted before fabrication to gain an average estimate idea regarding the cost involved in manufacturing the model. As most parts were reused or remanufactured items, the cost was comparatively less (Table 2).

5 Fabrication

The parts used for the fabrication of the model were mostly remanufactured items. Mild steel materials like hollow section, angle, and rectangular bar obtained from recycling center were used to build the overall structure of the model. Abrasive saw cutter was used to cut the materials according to the required dimensions. Arc welding was used to join the parts together into one frame. The base frame was built and three tyres were connected to the base. The two electric hub motors that were used to drive the vehicle were attached at the back portion of the base frame. The hub motors used in our model were previously used in electric scooters and these were obtained from the recycling center. The two hub motors were connected to the driving handle at the front of the vehicle which is at the height of the disabled in a wheelchair. The body of the vehicle was made using the remanufactured cabin from an Ashok Leyland lorry obtained from a recycling center. The motors were powered by four 12 V lead-acid batteries placed inside the vehicle. Solar panels were attached to the vehicle top and were connected to the batteries for charging. As the vehicle mostly uses solar-powered electric energy that creates no pollution, the fabricated model is environmental friendly. This could increase the free mobility of the disabled (Figs. 5, 6 and 7).

Fig. 5 Hollow section, angle, rectangular bar, respectively

Fig. 6 Hub motor, Lithium-ion batteries, respectively

5.1 Advantages

- Eco friendly design
- Ergonomic design
- Sustainable design
- Cheap and economic design.

5.2 Disadvantage

- Poor aesthetics.

Fig. 7 Fabricated model

6 Conclusion

It is true that everybody should have equal access to all resources and facilities irrespective of their physical well-being. All governments and organisations should work towards creation of such an ecosystem, where the disabled have equal access to all facilities like others. Transportation is one of the main problem faced by the disabled to find their own livelihood. Through this research work, we have found a solution for their transportation problems. The design was made using CREO and analysed using ANSYS software. The designed model after analysis was fabricated using recycled and remanufactured items. The cost analysis showed that, if recycled and remanufactured parts are used, the cost of materials can be decreased and thus the overall fabrication cost can be drastically reduced to make the vehicles cheap and more affordable to common people. The vehicle was designed to be solar-powered, so that it is not dependent on conventional fossil fuels that create environmental pollution. As a future scope, an aesthetically better design can be made without compromising the safety of passengers.

References

1. United Nations (2007) Accessibility; a guiding principle of the convention
2. Jones P (2011) Developing and applying interactive visual tools to enhance stakeholder engagement in accessibility planning for mobility disadvantaged groups. Res Transp Bus Manag 2:29–41
3. Griffin K W (2000) Building type basic for transit facilities
4. Henry T (2009) Policy on persons with disabilities, Office of the Prime Minister (Social Services Delivery)
5. Habert G, Blank, T (1992) Building design for handicapped and aged person. McGraw-Hill, Inc., United States
6. Hesla, B (2008) We have human rights. Harvard Project on Disability. Harvard
7. De Vos J, Schwanen T, van Acker V, Witlox F (2013) Travel and subjective wellbeing: a focus on findings, methods and future research needs. Transp Rev 33(4):421–442
8. Ettema D, Gärling T, Olsson L, Friman M (2010) Out-of-home activities, daily travel, and subjective well-being. Transp Res A 44:723–732
9. Jones P, Lucas K (2012) The social consequences of transport decision-making: clarifying concepts, synthesising knowledge and assessing implications. J Transp Geogr 21:4–16
10. Rye T, Mykura W (2009) Concessionary bus fares for older people in Scotland—are they achieving their objectives? J Transp Geogr 17(6):451–456
11. Levitas R, Pantazis C, Fahmy E, Gordon D, Lloyd E, Patsios D (2007) The multidimensional analysis of social exclusion. A research report for the social exclusion task force. University of Bristol, Bristol

Techno-Commercial Use of Silicon Carbide in Place of Ferrosilicon

Himanshu Shekhar Mishra, Rina Sahu, D. S. Padan, and Uttam Ganguli

1 Introduction

Generally grey iron is made from steel scrap, pig iron, foundry return, carburizers and ferrosilicon [1]. Using pig iron in grey iron making has advantages of development of more nucleation sites for graphite flakes [2]. If it is produced from steel scrap, nucleation sites are more but fade easily due to superheating of metals. If grey iron is produced using only steel scrap and ferrosilicon is added as silicon carriers. It acts as a pre-inoculator to the iron produced but it dissolves easily [3]. When steel scrap is used as base charge and foundry return is added in to make grey iron and silicon carbide is given in melt acts as a pre inoculator and dissolves slowly in melts. It develops nucleation sites more number in comparison to ferrosilicon [4, 5]. Silicon carbide is added in steel scrap initially before melting otherwise it would not dissolve in solution and efficiency decreases. Ferrosilicon is added after complete melt [6, 7].

Mechanical properties (tensile strength and hardness) of grey iron is improved after adding silicon carbide than ferrosilicon. Microstructural studies also reveal the flake size coarseness, uniformly distributed and 100% A type graphite whereas ferrosilicon in melt gives graphite flakes finer in size [8, 9], less uniformly distributed

H. S. Mishra (✉) · D. S. Padan · U. Ganguli
Foundry Divison, Tata Motors Ltd. Jamshedpur, Jamshedpur, Jharkhand 831010, India
e-mail: himanshu@tatamotors.com

D. S. Padan
e-mail: devinder.padan@tatamotors.com

U. Ganguli
e-mail: uttam.ganguli@tatamotors.com

H. S. Mishra · R. Sahu
Department of Metallurgical & Materials Engineering, NIT, Jamshedpur, Jharkhand 831014, India
e-mail: reenasahu.met@nitjsr.ac.in

R. Prasad et al. (eds.), *Advancement in Materials Processing Technology*,
Springer Proceedings in Materials 12,
https://doi.org/10.1007/978-981-16-3297-6_19

gives mostly A type with some D type also, using silicon carbide in making grey iron generates revenue in comparison to ferrosilicon [10, 11].

Silicon carbide is produced by the reduction of silica sand with carburizers (petroleum coke) in a cylindrical electric resistance furnace. It contains 70% silicon and 30% carbon by weight percentages. There are three layers formed. One which is accumulated around centre core graphite is 98% silica, second layer which is porous is generally used as raw material for making grey iron/ductile iron, third layer is white crust in color is known as siloksyskon. Silicon carbide material colour is varying with the presence of impurities present in it [12, 13].

This paper will elaborate on efficiency of making grey iron with silicon carbide with weight percentage of 50 and 100% both and also normal making of iron with the help of ferrosilicon only [14]. Method and steps of putting silicon carbide in furnace, its dissolution has been discussed. Samples have been made to know the improvement in tensile and hardness value. Microstructural studies have also been done for phases and types and distribution of graphite flakes in the matrix. Savings from making grey iron through silicon carbide route and ferrosilicon route has been discussed with comparison.

2 Material and Experimental Procedure

Steel scrap base charge is used for making grey iron. Steel scrap is used around 60% and foundry return 40% of the total capacity (5MT) of medium frequency furnace (Tables 1 and 2). Carbon percentage is increased by adding carburizers (petroleum coke) and silicon percentage is increased by adding ferrosilicon. Induction furnace is charged with steel scrap and petroleum coke and power is kept on at slowly initially and led upto 3800 KW. After melting of scrap, rest steel scrap and petroleum coke are also fed. After complete dissolution of scrap, foundry return and ferrosilicon (Table 3) is added one after another. Metal is heated and slag is removed. Sample for spectro analysis and wedge sample for chill depth are taken. Metal is superheated upto 1500–1510 °C before pouring into ladle. Three types of metals have been made. One with steel scrap base charge, foundry return, ferrosilicon and carburizers, second with steel scrap, foundry return and 50% silicon carbide and third with steel scrap,

Table 1 Steel scrap composition

% Carbon	% Mn	% Si	% S	% P
0.4	0.60	0.3	0.05	0.05

Table 2 Foundry return composition

% C	% Mn	% Si	% S	% P
3.35	0.60	2.00	0.05	0.05

Table 3 Ferrosilicon composition

% Si	% Fe	Efficiency
65	Rest	70%

Table 4 Silicon carbide composition

% Si	% C	Efficiency
70	30	85%

Table 5 Final chemistry of metal

Heat no	%C	%Si	%Mn	%P	%S	%Cr	%Cu	%Sn
PPA1 without SiC	3.35	1.96	0.63	0.046	0.058	0.05	0.43	0.043
PPA2 with SiC50%	3.35	1.94	0.63	0.046	0.059	0.05	0.47	0.043
PPA3 with SiC 100%	3.35	1.97	0.63	0.046	0.059	0.05	0.45	0.043

foundry return and 100% silicon carbide had been made (Table 4). Metal chemistry from all the three routes have been presented in Table 5. Three test bar samples had been made from each route metal. Samples were machined for making samples for tensile testing under universal testing machine. Samples were cut from tensile test pieces for making microstructural samples for matrix studies.

3 Results

Samples for study microstructural properties had been made ready by cutting from the broken pieces of tensile samples. It was cut to size on cutting machine. Uneven portion of samples was made even by rubbing on running belt conveyor of grit size (P 600). Polishing was done on emery paper of size 80 GI5.

Bar samples had been made from three different metals shown in Table 5. Bar sample was of size (300 mm height and 30 mm diameter) and it was made of sand mixed with resin binder. Before pouring into bar mould ferrosilicon 0.3% of size (. −18 +50) had been put. Bar mould was broken after complete cooling. These samples had been machined as per dimension shown in Fig. 1.

Tensile bars were tested under universal tensile testing machine. For each type of metal, three tensile samples had been made and results of these have been presented in Table 6.

Hardness samples had been made from the broken tensile samples and five readings had been taken on Brinell hardness testing machine. Hardness results has been shown below in Table 7.

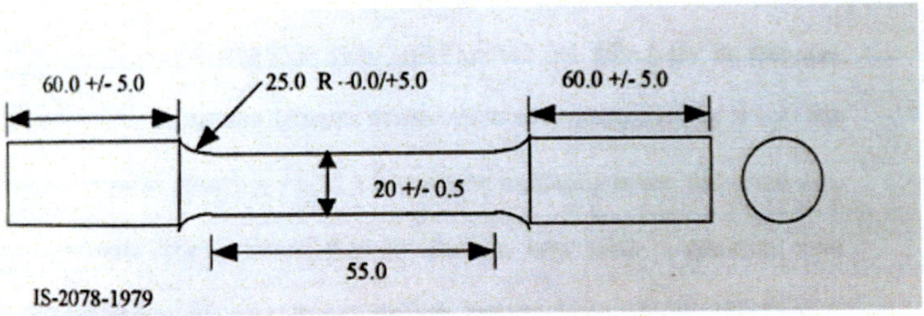

Fig. 1 Line diagram for tensile testing specimen for grey cast iron

Table 6 Tensile strength

Sample no	1	2	3
UTS1 without SiC	235.90	237.00	236.54
UTS2 with SiC 50%	255.00	255.50	257.70
UTS3 with SiC 100%	263.50	262.79	263.45

Table 7 Hardness

Sample1 witout SiC	212	212	215	214	212
Sample2 with SiC 50%	208	210	209	211	210
Sample3 with SiC 100%	206	200	209	198	202

Wedge chill test had been done for the three melt. Chill depth came shown below in Table 8.

Cost calculation of the three melts were done as per given below Table 9.

Table 8 Chill depth

Without SiC	6 mm	7 mm
With SiC 50%	5.5 mm	5 mm
With SiC 100%	4 mm	4 mm

Table 9 Cost calculation for three trial (Cost of CPC-Rs 24/Kg, Fesi-Rs.86/Kg, SiC-Rs.84/Kg.)

Various headings ↓	Trails →	Without SiC	With SiC 50%	With SiC 100%
CPC (Kg.)		150	135	135
FeSi (Kg.)		100	40	0
SiC (Kg.)		0	50	100
Total CPC cost		3600	3240	3240
Total FeSi cost		8600	3440	0
Total SiC cost		0	4200	8400
Total alloy cost/heat		12200	10880	11640
RS/MT of liquid metal		1805	1615	1733
Rs Savings/MT		0	189	71

4 Discussion

Microscope studies show that the sample prepared from the grey cast iron treated with fully FeSi (ferrosilicon) graphite flake is finer, uniformly distributed which is A type graphite 99% with less 1% D type graphite flake in without etch sample shown in Fig. 2a. As the melt is treated with Silicon carbide (SiC) and sample was seen under microscope shows graphite flake which is A type graphite flake with no other type. The graphite flake length becomes shorter that is bulky type, uniformly distributed in the matrix. Ferrosilicon (FeSi) is added in the melt after slag removal from complete melting of the steel scrap and foundry return. So it is added at higher temperature. Nuclei are formed in large numbers but dissolve easily that's why late inoculation is done in the metal before pouring into mould. Undercooling is more, so the time for growth of A type graphite is less and hence some D type is produced. But in case of melt is treated with SiC (silicon carbide) in Figs. 3a and 4a, under cooling is less and hence A type graphite growth time is more comparison to FeSi treated melt. Since SiC is given in steel scrap at the start of melt. It takes time to dissolve. Dissolution rate is very slow comparison to Fesi. As per literature,silicon

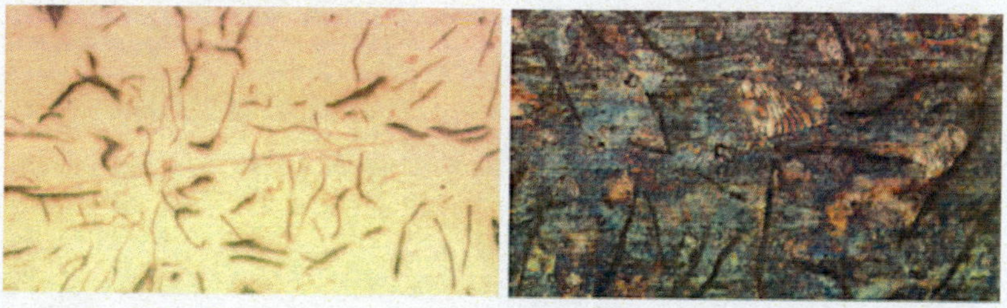

Fig. 2 **a** FeSi melt without etch at 100X magnification, **b** Fesi melt with etch at 100X magnification

carbide acts as pre-innoculator in the melt. Nuclei formation is more and stays upto last before pouring into mould. Late inoculation is required but of less quantity than to Fesi treated melt. For the same pouring temperature, liquidus temperature and eutectic temperature is increased and less undercooling is noted that is suitable for A type graphite but in case of ferrosilicon treated melt, both liquidus and eutectic temperature is less and more under cooling found which does not give time to form A type graphite fully.

Tensile strength is less incase of SiC treated sample as the matrix contains pearlite (Figs. 2a and 2b) and almost no ferrite whereas Fesi treated melt have some ferrite in Fig. 3b.Graphite fineness reduces in case of SiC-treated sample (Figs. 3a and 4a).

Hardness is also increased in case of SiC treated metal (Fig. 3a and 4a). Hardness is dependent on chemical composition of the melt, graphite morphology (size, distribution and shape) and on matrix.

Chill depth is low due to less undercooling and have pre inoculator effect in case of SiC treated metal.

Cost of calculating liquid metal from all three routes presented in Table 9 and savings per MT liquid metal by using 50% SiC and 0% ferrosilicon is Rs 189.

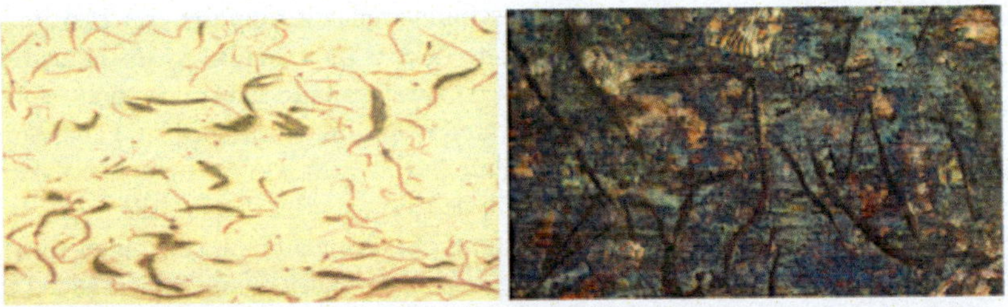

Fig. 3 **a** SiC 50% melt without etch at 100X magnification, **b** SiC 50% melt with etch at 100X magnification

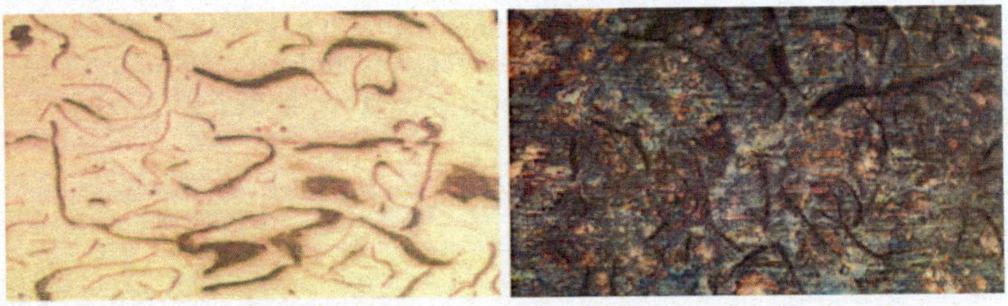

Fig. 4 **a** SiC 100% melt without etch at 100X magnification, **b** SiC 100% melt with etch at 100X magnification

5 Conclusions

Three experiments have been done first one the metal is made completely with ferrosilicon treatment, second one with 50% ferrosilicon and 50% SiC and third one with 100% SiC. Based on the above discussion, following points have been noted which is given below:

1. SiC addition is done initially in the steel scrap for complete dissolution in melt otherwise it would not melt completely if ferrosilicon process is followed. SiC efficiency will be low even after providing more time for dissolution
2. SiC acts as a pre-inoculator which gives A type of graphite in the microstructure whereas FeSi treated melt, undercooling is more and hence some D type graphite along with A type is produced.
3. Tensile and hardness is more incase of SiC treated metal than to FeSi treated melt.
4. Wedge chill depth is less in case of SiC treated melt due to pre inoculator effect and less undercooling.
5. Monetry savings is coming to Rs 189 per MT of liquid metal compared to ferrosilicon based melt.

Acknowledgements We would like to thank Foundry Division, Tata motors for allowing us to do trial on Silicon carbide for knowing its metallurgical effect as well as cost-effectiveness compared to ferrosilicon-based melt.

References

1. Janerka K, Bartocha D, Jezierski J, Szajnar J (2010) The effect of the carburizer quality on the cast iron recarburization process. Foundry Trade J Int 184(3679):282–286
2. Edalati K, Akhlaghi F, Nilli-Ahmadabadi M (2005) Influence of SiC and FeSi addition on the characteristics of gray cast iron melts poured at different temperatures. Department of Metallurgy and Materials Engineering, Faculty of Engineering, University of Tehran, P.O. Box 1136–4563, Tehran, Iran. J Mater Process Technol 160, 183–187
3. Schubert WD, Ta AT, Kahr G, Benecke Th, Lux B (1985) Institute for chem. Techno anorg. Stoffe der Techn. Univ. Wien. Influence of SiC additions on the microstructure of gray cast iron. Mat Res Soc Sym Proc 34
4. Stojczew A, Janerka K, Jezierski, Szajnar J, Pawlyta M (2014) Department of Foundry Engineering, Silesian University of Technology, Towarowa, 44–100 Gliwice, Poland. Melting of grey cast iron based on steel scrap using silicon carbide. Archiv Foundry Eng 14(3). ISSN (1897-3310)
5. Heine RW, Loper CR, Rosenthal PC (1976) Principal of metal casting, TMH ed. McGraw-Hill
6. Onsoien MI, Skaland T (2001) Preconditioning of gray iron melts using ferrosiliconor silicon carbide. Am Foundry Soc
7. Riposan I, Chisamera M (1993) Use of silicon carbidein cast iron. Kohassz 126(9):328–330
8. Benecke T, Ta AT, Kahr, Schubert WD, Lux B (1987) Dissolution behaviour and pre-innoculation effect of SiC in gray cast iron. Giesserei 74(10):301–306

9. Copi KW (1993) The use of silicon carbide in today's iron foundry. In: Proceedings of the 51st electric Furnace Conference, pp 35–38. Washington, DC, USA
10. Rivera GL, Boeri RE, Sikora JA (2004) Solidification of cast iron. Scripta Mater 50:331–335
11. Caspers KW (1972) Use of silicon carbide in main-frequency induction furnace. Gieseserei 59(18):656–659
12. Mizoguchi T, Perepezko JH Nucleation behaviour during solidification of cast ironat high undercooling. Mater Sci Eng A 226–228
13. Hayama F, Kayama N, Toyoda Y (1970) Effect of siicon carbide addition on the machinabilityof cast iron. Imono 42(3):195–199
14. Janerka K, Jezierski J, Szajnar J (2011) The evaluation of effectiveness of liquid cast iron recarburization process by using different carburizers and methods of recarburization. In: Proceedings of 20th anniversary international conference on metallurgy and materials METAL, 2011, Brno, Czech Republic

Modified Die Lubrication System for Forging Wheel and Axle Plant

S. K. Jha, D. K. Jain, S. De, M. Chakraborty, and D. Karmakar

1 Introduction

The railway wheels are manufactured at Durgapur Steel Plant, primarily to meet the requirements of Indian Railways for Broad Gauge (BG) coaches and Broad Gauge locomotives. The wheels produced are supplied as per the Indian Railway Specification R-19-93 for BG coaches and R-34-99 for BG locomotives.

Wheel manufacturing process involves casting in cylindrical moulds in the steel melting shop. Afterwards these ingots are cut into blocks having requisite weights as per type of wheel to be manufactured. These blocks are heated in computer-controlled rotary hearth furnace to around 1250–1300 °C. These blocks after descaling are pressed in 63/12 MN oil hydraulic press. Hydraulic press consists of 2 top dies and 3 bottom dies, which are placed horizontally. Forming process is carried out in closed dies.

Forging is carried out in 3 steps: (i) Upsetting, (ii) Forming and (iii) Punching. Die positions and pressing operation are ensured by fixed programmes. Movement of both the top tool frame and bottom sliding table are longitudinal, whereas top dies moves in vertical downward direction during upsetting, forming and punching operation. The centering device ensures placement of the wheel blanks at the centre of the press.

For designing any die lubrication system, understanding its effect on entire forging operation is essential. When die designers select a tool steel, they usually do not focus

S. K. Jha (✉) · D. K. Jain · D. Karmakar
Reasech & Development Centre for Iron & Steel, SAIL, Ranchi, India

D. K. Jain
e-mail: dkjain@sail.in

S. De · M. Chakraborty
Durgapur Steel Plant, SAIL, Durgapur, India
e-mail: m.chakraborty@sail.in

R. Prasad et al. (eds.), *Advancement in Materials Processing Technology*,
Springer Proceedings in Materials 12,
https://doi.org/10.1007/978-981-16-3297-6_20

on whether the fluid actuation is integrated with the press controls [1]. The principal reason for applying fluids and lubricants in forming processes is to reduce friction and remove or dissipate heat [2]. The heat and friction are generated at the point where the tool and the work piece meet that is where the shaping, forming, bending, drawing, cutting, or punching takes place. The goal with die lubrication is to make sure an appropriate film of lubricant is at the interface between the tool and the workpiece to minimize or eliminate damage to the die and markings on the part [3].

The goal with fluid application is to find the best possible method to deliver the desired amount of lubricant to the area in the die, where it is required and where it will be the most effective [4]. Advantages of direct die lubrication include prolonged tool life, die protection, and precision routing of the fluid to its destination [5]. The primary difficulty with direct die lubrication is that the fluid sometimes is applied where it is not needed. This increases consumption and requires appropriate attention to deal with the excess fluid. Normally in close die forging 8–12% lubricant is used [6]. Important improvements also occur when die life is prolonged. Downtime caused by die changeover and die maintenance can be reduced when lubricants reach critical areas in the most efficient manner [7]. In a Hot Forging process die lubrication is the key for quality and productivity. Lubrication is dependent on the right selection of parameters such as: Lubricant and application types, dilution percentage, even when fully automated systems are employed [8].

There is a relationship between die wear, die strength and friction factor. For proper prediction of die life actual die temperature, coefficient of friction and heat transfer coefficients are important. During upsetting and forming operation metal flow depends on surface condition of dies and friction between dies and work piece [9].

At Wheel and Axle plant die lubrication was carried out using graphite powder. After completion of upsetting, forming and punching operations, scale was removed from the dies by compressed air and then graphite powder was sprayed by shovel on die. By using this method most of the time graphite was not able to cover full surface area of die. Therefore by this method dies were not lubricated properly resulting in poor die surface and uneven metal flow during forming operation.

2 System Design

For improvement in die life and better dimensional accuracy, an automatic die lubrication system was designed (Fig. 1).

The system consists of Tank Assembly (SS Tank, Stirrer, Pump, Level Indicator), die lubricant spray unit, lubricant spraying header consisting of nozzles, hoses electrical panels, etc. Pneumatic pump having capacity of 60 ± 10% lpm at 5–7 bar is used for spraying lubricant on dies. The system is operated based on timer and is interfaced with PLC controls of the press. Interfacing signals from press are taken to the lubrication system. Lubrication system is operated by two control panels, for manual and Auto mode operation. In mixing tank water soluble graphite is mixed with water.

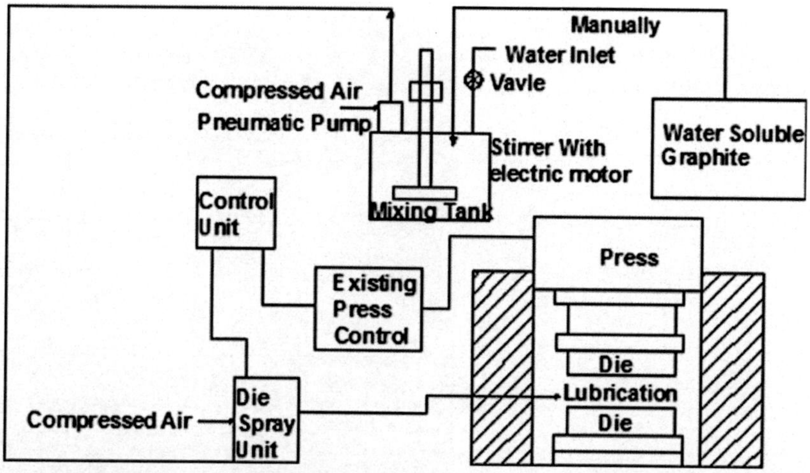

Fig. 1 Schematic drawing of die lubrication system

Quantity of lubrication and duration of spray was decided based on concentration of lubricant required and available time between finishing of operation and charging of subsequent wheel. Die lubrication start after completion of upsetting and forming process. Automated lubricating system was designed to work in both auto and as well as manual mode.

System was designed to lubricate top and bottom dies of upsetting and forming press. In first operation upsetting and forming dies are cleaned of scales by compressed air at a pressure of 5 bar. After that, when dies are stationary position following upsetting operation, top and bottom upsetting dies are lubricated. Afterwards after completion of forming operation top and bottom forming dies are lubricated. Orientation of spray was fixed in such a way to lubricate punch also after finishing of piercing operation, as many a time sticker formation between punch and work piece takes place (Figs. 2 and 3).

During operation, work piece temperature is around 1200–1250 °C and dies are round 300–400 °C. In past, oil-based lubricant was used by handheld spraying gun for lubrication, but it resulted in thick fumes formation. During use of oil-based lubricant there were few instances of fire on dies. Based on these experiences water-based lubricant was selected.

Properties of lubricant used are shown in Table 1.

Lubricant is sprayed on the dies with the help of suitable headers. Automated system have sensor for auto-stopping in the event of malfunctioning of lubrication system or during sticker formation. In event of malfunctioning, system has selector switch to bypass die lubrication system for normal press operation.

The system is controlled by timer and is interfaced with PLC controls of the press. Lubrication system gets signals from press and dies lubrication start after completion of upsetting and forming process.

Fig. 2 Tank assembly for die lubrication

Fig. 3 Panel for lubrication system

Table 1 Properties of lubricant

Specific gravity at 150 C	1.1–1.4
pH	7.5–10.0
Active solid contents	40% ± 5%
Uniform lube film formation	400 °C
Density	1000–1200 kg/m^3

Die lubrication system was installed and commissioned. During trial various adjustments were carried out to take maximum benefit of this system. Orientation of nozzle was adjusted so that it covers full area of dies. Initially during trials punch was not moving freely in and out of work piece. Orientation of nozzle was changed so that lubricant is sprayed on punch also. To control cost of lubricant initially lubrication was carried out with 5% concentration. But during trial 4–5 numbers of sticker (between work piece and dies) formation took place in a shift. Afterwards lubrication concentration was increased to 7%, sticker formation was reduced but still 1–2 sticker formation took place in a shift. Subsequently, lubricant was diluted to 10% and then problem of sticker formation was almost eliminated. Lubrication with concentration of 12% was also tried out but during trial fumes from dies increased and thereafter trial was resumed with 10% concentration.

Spray time of the nozzle was also adjusted during trial period. Duration of spray on upsetting die was reduced to 2 s and that on forming dies was increased to 4 s. This was because in upsetting dies contours and cavities are not present but in forming dies cavities are present. Therefore chances of sticker formation are higher in forming dies.

3 Results and Discussion

Forged-rolled and heat-treated wheels are machined before online testing. At machining stage, any deviation in dimension is visible which is not possible before that. At any location, if no/less material is available for machining then wheel is rejected and it is known as wash out. These defects are at machining stage but genesis of these defects are forming stage. Wash out of wheels is mainly on tread and web portion. Major reasons for wash out are eccentricity, improper material flow and wear out dies and rolls. Reason for improper material flow is lack of lubrication on dies which results in constrained flow of metal during upsetting and forming. Purpose of die lubrication system was to reduce/minimize rejections of account of wash out at machining stage by better flow of material during upsetting and forming operations.

Implementation of die lubrication system resulted in reduction in rejections on account of Mill (forging and rolling) at machining stage by more than 2%. Lubrication system impacts rejections as metal flow improves by using efficient lubrication thereby reduction in wash out.

Fig. 4 Physical condition of dies before and after lubrication

Earlier there were instances of sticker formation (between work piece and dies) during forming process (around 6–8 per shift) because of lack of lubrication but after use of lubrication system no sticker formation has taken place. Apparently physical condition of dies was also improved (Fig. 4).

4 Conclusions

- Metal flow during upsetting and forming improves by proper use of lubricant.
- Use of proper lubrication reduces chances of sticker formation during forging operation.
- Water-based die lubricant should be used, as use of oil-based lubricant may result in environmental issues along with chances of fire during forging.
- For efficient lubrication to impart expected performance angle, pressure, distance and proper atomization are essential.
- Proper concentration of die lubrication should be used. In wheel forging (close die forging) 8–12% concentration of lubricant is desired.
- Nozzle orientation should be proper and full surface of dies should be covered. Extra care is to be taken so that lubricant is sprayed on punch also.

Acknowledgements The authors humbly acknowledge concerted efforts by all the task force members of Wheel and Axle Shop, Durgapur Steel Plant and Research and Development Centre Iron and Steel, Steel Authority of India Limited, Ranchi for their cooperation during execution of this work.

References

1. Gustavo G, Schiuma E (2014) Die lubricants for hot forging
2. Hancox A, Hancox D, Haigh G (1994) Codes of best forging practice, die lubrication in hot forging, June. British & Midland Forging Ltd., Britain
3. Ideal Temperature for Application of Lubricants. http://www.renite.com/forgelub.htm
4. Kumar M (2013) Institutes Class Literature, National Institute of Foundry and Forge Technology, Associate Professor, Department of Forge, Ranchi
5. Yang L, Shivpuri R (2006) Spreading behavior of water based graphite lubricants on hot die surfaces. Industrial, welding and systems engineering. Ann CIRP 55(1). The Ohio State University, Columbus, Ohio, USA
6. Behrens BA, Barnert L, Huskic A (2005) Alternative techniques to reduce die wear—hard coating or ceramic? Annals of the German Academic Society for Production Engineering (WGP)
7. Mitamura K, Fujikawa S Application of boride cermet in warm forging dies. Nissan Mortor Co., Ltd.
8. Lange K (1985) Handbook of metal forming. Society of Manufacturing Engineers, ISBN 0-87263-457-4
9. Sawamura M, Yogo Y, Kondo S, Tanaka T, Nakanishi K, Suzuki T, Watanabe A (2005) Estimation of spray lubrication and die temperature for die wear life prediction in hot forging. R & D Rev Toyota CRDL 40(1)

Investigation of Effects of Different Heat Treatment Cycles Combined with Quenching Partitioning Treatment on Mechanical Properties of High Carbon Spring Steel

Prakash G. Ranaware, Manoj J. Rathod, and Venkatesh G. Dhorde

1 Introduction

In modern world, the primary applications of steel are in automobile industries and in civil constructions. The main requirements for these steels are suitable combination of high strength and good formability. In past few decades, the steel industry has developed many types of steels termed as advanced high strength steels (AHSS) such as dual-phase (DP) steels, transformation-induced plasticity (TRIP) steels, complex phase (CP) steels, high strength low alloy (HSLA) steel and few other types of steel which possess combination of good strength and ductility. This is chiefly achieved by controlling microstructural features of steel through controlled thermo-mechanical processing and heat treatment. All over the world continuous attempts are being made to get the optimized combination of strength and ductility at reduced cost [1]. Bhadeshia [2], Bhadeshia and Edmonds [3], Kozeschnik and Bhadeshia [4], Wang and Zwaag [5] and Pereloma et al. [6] showed that the amount of carbon partitioned in retained austenite (RA) during bainitic transformation, i.e., during austempering enhances with increase in the carbon content in the parent austenite phase. However, this enrichment of carbon in austenitic phase is possible only if cementite precipitation is suppressed during bainitic transformation. For this purpose, extra silicon is generally added in the steel composition as silicon, strong graphitizer, inhibit precipitation of cementite. It is reported by number of researchers that in low or medium carbon high silicon steels, optimal combination of ductility and strength can be achieved through austempering by isothermal transformation at the temperature of 350 °C Varshney et al. [1, 7–9].

Quenching and partitioning (Q&P) is newly developed heat treatment technique which involves quenching of steel after austenitizing at the temperature between

P. G. Ranaware (✉) · M. J. Rathod · V. G. Dhorde
Department of Metallurgy and Materials Science, College of Engineering Pune, Pune, India
e-mail: ranawarep.meta@coep.ac.in

R. Prasad et al. (eds.), *Advancement in Materials Processing Technology*,
Springer Proceedings in Materials 12,
https://doi.org/10.1007/978-981-16-3297-6_21

Martensitic start (M_s) and Martensitic finish (M_f) temperature, holding there for few seconds and followed by isothermal holding at slightly higher temperature (Partitioning) before final quenching [10]. The later step enriches untransformed austenite in carbon which helps to stabilize austenite. Final microstructure contains mixture of martensite and austenite which provides combination of strength and ductility. Continuous cooling from austempering temperature with controlled cooling rate can give multiphase structure such as ferrite, bainite, martensite and austenite by controlling cooling rate and heat treatment processing route. This simple heat treatment can also be used to obtain desired combination of mechanical properties. A modified heat treatment was developed by [11] called quenching, partition and quenching (QPT) process produces similar results.

In this work, a study of effect of combination of heat treatment cycles on mechanical properties of high carbon spring steel was done which contains very low silicon. These steels have important applications in manufacturing automobile parts and machining tools. In general, high carbon spring steels possess high tensile strength (1500 MPa) and yield strength (1100 MPa) but less ductility (less than 10%). These steels show less toughness compared to low and medium carbon structural steels. Objective of this work was to find out combination of heat treatment cycle which can produce suitable combination of mechanical properties. Thus, different combination of heat treatment cycles such as continuous cooling, austempering, quenching and partitioning, QPT were conducted on high carbon spring steel. After the heat treatments, characterization and testing of specimens were done by hardness testing, tensile testing, optical microscopy, SEM and EBSD.

2 Materials and Methods

Chemical composition of spring steel used for this study is shown in Table 1. Cylindrical specimens of 6 mm diameter and 300 mm height were cut from a single rod. All specimens were annealed at 950 °C before subjecting them to heat treatment cycles. Before conducting heat treatments, critical temperatures Ac_1, Ac_3, M_s, and B_s were calculated using empirical relations given in reference Varshney et al. [1, 7] which comes out for the composition given in Table 1 as 850, 740, 248, and 483 °C, respectively.

Two heat treatments cycles (namely HTC-1 and HTC-2) were applied to specimens. HTC-1 is combination of Q&P and austempering heat treatment cycles. HTC-2 combines continuous cooling (CC), Q&P and modified QPT cycles. In HTC-1 specimens were austenitized at 870 °C for 10 min and then quenching was carried at 230 °C

Table 1 Composition of steel used in experimentation

Element	C	Si	Mn	Cr	Mo	Ni	S	P	Fe
(Wt%)	0.587	1.336	0.905	0.21	0.071	0.022	0.028	0.034	Remaining

and partition was done for 10 s. The specimens were then immediately austempered at three temperatures namely 300, 350 and 400 °C for 2 h and at the end all specimens were quenched in water. In HTC-2 specimens were autenitized at 1040 °C for 10 min and then continuously cooled in air for two different times namely 13 and 18 s to achieve different amount of intercritical ferrite and austenite with different enrichment of carbon. After continuous cooling specimens were quenched at 50 °C and partitioned for 10 s. After Q&P, modified QPT cycle was applied which involved austempering at two different temperatures (300 and 350 °C) for 90 min which was followed by tempering for three different temperatures (400, 450, 500 °C) for 90 min.

After heat treatments, all the specimens were tested for hardness on Vickers hardness tester and for tensile test on UTM machine (MUTC-20 Series Universal Motion INC). Optical microscopy of the specimens was done on SuXma-Met-1 microscope at 100X and 500X magnifications and the micrographs were processed using Conation Micrograph Analysis Software. Specimens were also then analyzed on Scanning electron microscope of SIGFA-HV whose resolution is 1.3 nm (20 kV). Accelerating voltage of SEM is 0.1–30 kV. The EBSD scans were performed on the cross-section of a specimen. EBSD data was analyzed by using TSL-OIM 7.1® data analysis software.

3 Results and Discussion

After performing the heat treatment cycles as discussed in previous section, results are presented and discussed in this section.

3.1 Heat Treatment Cycle 1 (HTC-1)

Figure 1 shows optical microstructures for the specimens subjected to HTC-1. Figures show martensitic, bainite along with retained austenite. No traces of carbides were found in optical microscopy which indicates that they may be extremely fine which are not able to resolve by optical microscopy. Volume fractions of retained austenite were estimated by using image analysis software and amount of retained austenite was found to vary from 18 to 22%. Figure 1d shows SEM image of specimen after austempering at 300 °C. The image also clearly shows martensite, bainite and retained austenite and also absence of carbides.

Mechanical properties after HTC-1 are tabulated in Table 2. Highest hardness, YS and UTS values were observed for the specimen austempered at 300 °C. Specimen austempered at 350 °C showed slightly higher ductility than specimen austempered at 300 °C. However, product of UTS and %elongation was observed to be highest for specimen austempered at 300 °C. This is due to presence of higher volume fraction of bainite in the structure along with martensite and presence of retained austenite. It has been established in the literature that amount of bainite decreases with increase

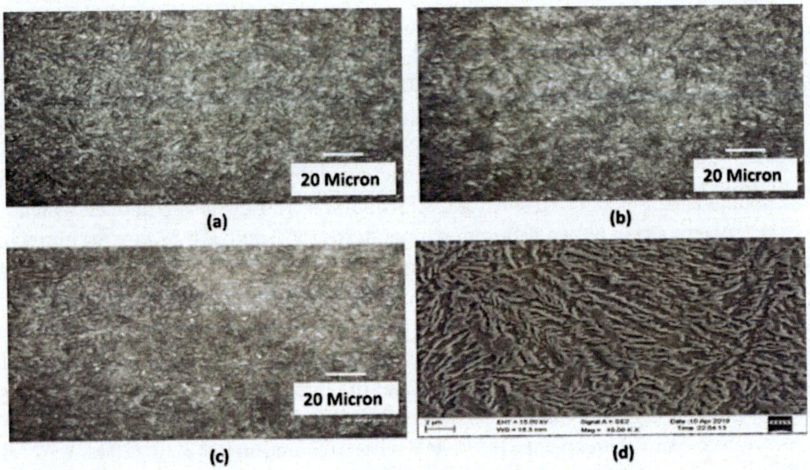

Fig. 1 Optical microstructure of specimens subjected to HTC1 austempered at **a** 300 °C **b** 350 °C **c** 400 °C **d** SEM image of specimen subjected to HTC1 austempered at 300 °C

Table 2 Mechanical properties for specimens subjected to HTC1

S. N	Austempering temperature (°C)	Hardness (HV)	YS (MPa)	UTS (MPa)	%El	UTS*%El
1	300	660	1512	1748	10.50	18,354
2	350	587	1211	1448	11.65	16,995
3	400	509	1057	1264	9.40	11,881

in austempering temperature [2]. Volume fraction of bainite is not estimated in this work.

3.2 *Heat Treatment Cycle 2 (HTC-2)*

Figure 2 shows microstructures of specimen subjected to HTC-2. Microstructures at all three austempering temperatures show presence of tempered martensite, bainite, ferrite and retained austenite. Ferrite is formed during intercritical cooling. All the microstructure shows no traces of carbides and also no peartlite was observed in the microstructures. Figure 2d shows SEM image for the specimen subjected to HTC-2 where it was austempered at 300 °C and finally tempered at 450 °C. In this micrograph tempered martensite, bainite, ferrite and retained austenite are visible.

Mechanical properties of the specimens subjected to HTC-2 are tabulated in Table 3. It is observed that the hardness value is maximum for the specimen continuously cooled for 13 s and austempered at 300 °C followed by tempering at 450 °C.

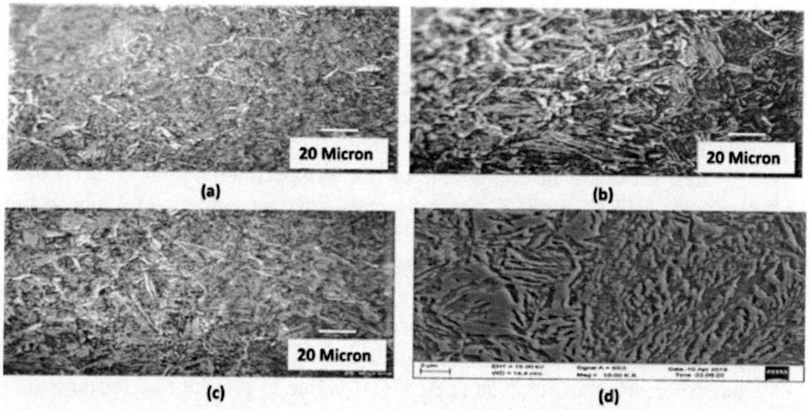

Fig. 2 Optical microstructure specimens treated with HTC-2 with continuous cooling for 13 s, quenching at temp. 50 °C, partitioning at 300 °C and tempered at **a** 400 °C **b** 450 °C **c** 500 °C **d** SEM image of specimen subjected to HTC-2 with similar conditions and tempered at 300 °C

Table 3 Mechanical properties of specimens subjected to HTC-2

S. No	Specimen name	Hardness (HV)	Yield strength (MPa)	UTS (MPa)	%El	UTS*%El
1	T13A300Q400	572	1354	1579	9.34	14,746
2	T13A300Q450	626	1264	1440	16.67	24,004
3	T13A300Q500	611	1092	1240	15	18,605
4	T13A350Q400	581	1101	1259	11.65	14,662
5	T13A350Q450	548	1107	1287	11.58	14,904
6	T13A350Q500	483	1118	1311	12	15,726
7	T18A300Q400	441	1318	1534	9.58	14,700
8	T18A300Q450	511	1136	1365	11.2	15,288
9	T18A300Q500	605	1022	1210	16.98	20,546
10	T18A350Q400	560	1463	1686	8.97	15,122
11	T18A350Q450	596	1440	1637	12	19,638
12	T18A350Q500	612	1115	1327	11.92	15,811

Nomenclature: T13A300Q400 means Continuous cooling for 13 s, austempering at 300 °C and tempering at 400 °C

Maximum tensile strength was observed for the specimen austempered at 350 °C and tempered at 400 °C after continuous cooling for 18 s. However, maximum ductility was observed for the specimen continuously cooled for 18 s, austempered at 300 °C and tempered at 500 °C. The highest product of UTS and % elongation was observed for specimen continuously cooled for 13 s, austempered at 300 °C and tempered at 450 °C. These results show interplay of phases like ferrite, martensite, tempered martensite, bainite and retained austenite. This calls for thorough investigation to ascertain role of each phase.

EBSD was done on the specimen which gave best combination of strength and ductility. This specimen was treated by HTC-2 for continuously cooling of 13 s, quenching at 50 °C and austempered at 300 °C and finally tempering at 450 °C. Figure 3 shows the inverse pole figure for the specimen. Figure 3 also shows volume fraction of FCC and BCC phase. Here retained austenite is FCC phase and ferrite, ferritic bainite and martensite constituted BCC phase. Surprising, the volume fraction of FCC phases is very low (0.006%). This shows that austempering at 300 °C and subsequent tempering at 450 °C, eliminate all the austenite. It also indicates that silicon in the specimen is not sufficient to stabilize the austenite by carbon enrichment. EBSD results indicate that the higher ductility values obtained in HTC-2 are not due to presence of retained austenite, it is due to presence of ferrite and tempered martensite.

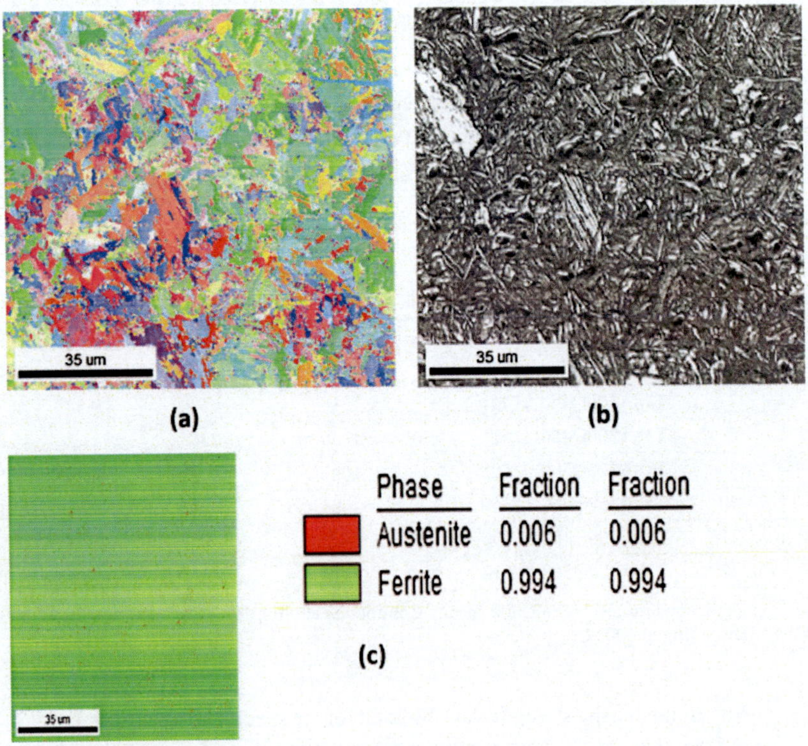

Fig. 3 EBSD analysis for the specimen treated with HTC-2 with nomenclature T13A300Q450 **a** Inverse pole figure **b** Image quality **c** Volume fraction of austenite and ferrite

4 Comparison Between Two Cycles

It is clear from previous sections, tensile properties like YS, UTS and %elongation are comparatively higher for HTC-2 than HTC-1. Products of UTS and % elongation are in general higher for HTC-2 than HTC-1. These results indicate that continuous cooling, Q&P and modified QPT generate superior results compared to combination of Q&P and austempering cycle. These results of best combination of mechanical strength and ductility are obtained in spring steel with low silicon content which indicates that even without retained austenite, it is possible to achieve strength and ductility combination with multiphase microstructure containing phases like ferrite, bainite, and tempered martenite.

5 Conclusions

After carrying out new heat treatment cycles comprising continuous cooling, quenching and partitioning, austempering, and QPT and analysing the results following conclusions are drawn.

1. Combination of heat treatments like CC, Q&P, austempering and QPT can produce steel with good combination of tensile strength and ductility.
2. Mechanical properties are superior for the specimens subjected to CC, Q&P, followed by modified QPT (HTC-2) compare to the specimens subjected to Q&P followed by austempering (HTC-1).
3. Product of tensile strength and % elongation is maximum for specimens treated with HTC-2 than specimens treated with HTC-1 which indicates that multiphase microstructure containing ferrite, bainite and tempered martensite can produce better combination of mechanical properties than specimens containing microstructure of martensite, bainite and retained austenite.
4. EBSD results show that the higher values of ductility are due to the presence of ferritic phase and tempered martensitic structure and not due to retained austenite for the specimens subjected to HTC-2.

References

1. Varshney A, Sangal S, Kundu S, Mondal K (2016) Super strong and highly ductile low alloy multiphase steels consisting of Bainite, Ferrite and retained Austenite. Mater Des 95:75–88
2. Bhadeshia H (2005) High performance Bainitic steels. Mater Sci Forum 500–501:63–74
3. Bhadeshia H, Edmonds DV (1983) Bainite in silicon steels: new composition–property approach part 2. Mater Sci 17:420–425
4. Kozeschnik E, Bhadeshia HK (2008) Influence of silicon on cementite precipitationin steels. Mater Sci Technol 24A:343–356

5. Wang J, Zwaag SV (2001) Stabilization mechanisms of retained austenite in transformation-induced plasticity steel. Metall Mater Trans A 32A:1527–1553
6. Pereloma EV, Gazder AA, Timokhina IB (2016) Retained austenite: transformation-induced plasticity. Encycl Iron Steel Their Alloys I Part A 3088–3103
7. Varshney A, Sangal S, Kundu S, Mondal K (2016) Superior work hardening behaviour of moderately high carbon low alloy super strong and ductile multiphase steels with dispersed retained austenite. Mater Des 99:439–448
8. Varshney A, Sangal S, Mondal K (2017) Exceptional work-hardening behavior of medium-carbon high-silicon low-alloy steels. Metall Mater Trans 48A:590–593
9. Varshney A, Sangal S, Pramanick AK, Mondal K (2017) Microstructural evidence of nnano-carbides in medium carbon high silicon multiphase steels. Mater Sci Eng A 708:237–247
10. Wang L, Speer JG (2013) Quenching and partitioning steel heat treatment. Metallograph Microstruct Anal 2:268–281
11. Zuyao X (2009) A brief introduction to Quenching-Partitioning-Tempering(Q-P-T) process. Heat Treat Met 34(6):1–8

Remarkable Effect of Graphene on the Properties of FeCoCrNi-Based High Entropy Alloy

Vaishali Poddar, Narendra Dhokey, Akshata Gole, and Rutuja Dongare

1 Introduction

High Entropy Alloys (HEAs) are a new effort in materials science and engineering. They exhibit exceptional properties and thus have potential applications in the transportation and energy sectors. HEAs are synthesized with an aim to develop structural materials and being a novel material, research is still in its rudimentary stages. The main steps of HEA development are design, evaluation and synthesis using one process route or a combination of process routes. There is extensive and thorough research on the hardness and wear resistance of HEAs synthesized by melting and casting and surface deposition techniques [1–3]. This research work aimed to develop HEA with an optimum hardness and wear resistance. Further, the study of HEAs synthesized by the powder metallurgy route is limited which serves as the prime motivation for this research work.

2 Experimental Procedure

The metal powders of cobalt, chromium, nickel, iron and graphene or niobium (grades, purity and particle size are mentioned in Table 1) were subjected to high energy ball milling in an in-house developed attritor mill (SS 304 vial equipped with water cooling system, 6 mm Φ SS 304 balls, 400 rpm, BPR = 10:1, Ar atmosphere). The amount of powders to be used was calculated stoichiometrically. The stoichiometry for graphene composition was Co:Cr:Fe:Ni:C::1:1:1:1:0.2 and for niobium composition Co:Cr:Fe:Ni:Nb::1:1:1:1:0.1.The milled powder was subjected to hot

V. Poddar (✉) · N. Dhokey · A. Gole · R. Dongare
Department of Metallurgy and Materials Science, College of Engineering, Pune 411 005, India
e-mail: poddarp.meta@coep.ac.in

R. Prasad et al. (eds.), *Advancement in Materials Processing Technology*,
Springer Proceedings in Materials 12,
https://doi.org/10.1007/978-981-16-3297-6_22

Table 1 The grades, purity and particle size of the powders used

Element	Grade	Purity (%)	Particle size
Cobalt	Metal basis	99	200–250 mesh
Chromium	Metal basis	99	200–250 mesh
Iron	Electrolytic grade	99	250–300 mesh
Nickel	Metal basis	99.5	150–200 mesh
Niobium	Metal basis	99.8	325 mesh
Graphene	Grey flakes	91–94% Carbon, 4–5% Oxygen, 0.1–0.3% Sulfur,0–1% Hydrogen	325–400 mesh

press sintering (HPS Equipment, Make-KEJETHERM, and Maximum Capacity: 8 Ton, 1673 K) in the graphite mould. The samples obtained from HPS were placed in a ceramic boat for heat treatment. The heat treatment was carried out in an inert argon atmosphere at the temperature of 1300 °C with a holding time of 4 h in each case. Laser particle size analysis (Model: LA-960, Make: Horiba) was performed for all the milled powders. The phase transformation after hot press sintering and heat treatment was characterized by X-ray diffraction (XRD) analysis. The XRD analysis was carried out on X'Pert PRO Analytical diffractometer (Model: Goniometer-PW3050/65) and X-Ray Diffractometer, D-8 Advance (Make- BRUKER AXS). Phase transition and microstructure evolution during the process were investigated using SEM–EDS (Make: CARL ZEISS, sigma model).

3 Results and Discussion

The enthalpy and Gibbs free energy of the selected material system contribute to its resultant entropy [4]. A semi-empirical approach called the Miedema Model was designed for $FeCoCrNiC_x$ and $FeCoCrNiNb_y$ HEA systems as shown in Fig. 1. It assisted to understand the phase formation in accordance with the thermodynamic calculations. The interpretations of these calculations gave an idea about the phase formations.

The thermodynamic reactions involved during the alloy formation are of prime importance while determining the phases formed. The solubility of one element in another is affected by various factors like mixing entropy, atomic size, electronegativity, etc. In the multicomponent system like HEA, and specifically the systems of our interest are represented in Table 2 including the interpretation with respect to each of the calculated values.

Where R = Universal Gas constant, x_i = mole fraction of the *ith* element, r_i = atomic radius of the *ith* element, VEC_i = Valence electron concentration (VEC) of the *ith* element, T_m = average value of the melting temperature, ΔH_{mix} = mixing enthalpy, X_i = electronegativity of the *ith* element, φ_i = coefficient of the element

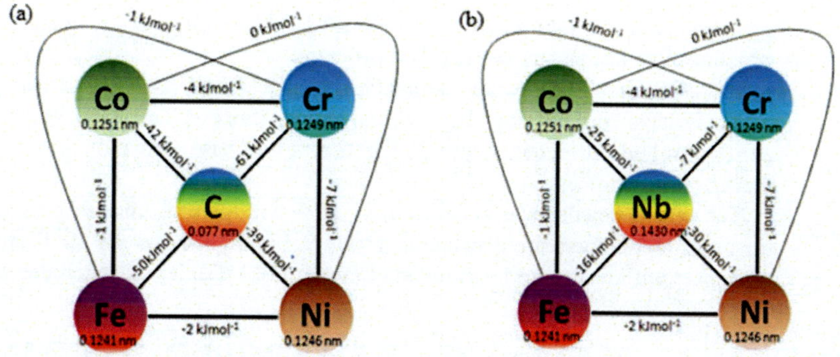

Fig. 1 Miedema model for **a** $CoCrFeNiC_x$ and **b** $CoCrFeNiNb_y$

Table 2 The thermodynamic requirements for the HEAs to form the solid solution: $CoCrFeNiC_{0.2}$ and $CoCrFeNiNb_{0.1}$

Sr. no	Parameter	Interpretation	$CoCrFeNiC_x$	$CoCrFeNiNb_y$
1	Mixing Entropy [4]: $\Delta S_{mix} = -R\sum_{i=1}(x_i ln x_i)$	If $\Delta S_{mix} > 1.61R = 13.38 Jmol^{-1}$, then a solid solution forms else partial solid solution will form	12.198 J mol^{-1}	12.114 J mol^{-1}
2	Size difference [4]: $\delta = 100\%\sqrt{\sum_i x_i(1 - r_i/\overline{r})^2}$, where $\overline{r} = \sum_i x_i r_i$	If $\delta < 6.6\%$ then a solid solution forms else partial solid solution is formed	5.92%	17.8%
3	Valence electron concentration[4]: $VEC = \sum_{x_i} x_i VEC_i$	If VEC < 7.5, then bcc structure else fcc structure	8.14	8.162
4	Mixing Entropy effect [5]:$\Omega = \frac{T_m \Delta S_{mix}}{\Delta H_{mix}}$	If $\Omega > 1.1$, then a solid solution forms	3.05	4.4
5	Electronegativity [5]: $\Delta X = \sqrt{c_i(\overline{X} - X_i)^2}$ where $\overline{X} = \sum_i x_i X_i$	If $\Delta X > 1.33$, then a solid solution forms else partial solid solution is formed	0.31	0.1
6	FCC-BCC Index [6]: (FBI $= \sum_{x_i} x_i \varphi_i$)	FBI = dependence of crystal structure of the alloy on the individual elements, 1 = fcc,–1 = bcc	-0.5	−0.27

with respect to its crystal structure; $\varphi_i = 1$, if fcc structure, $\varphi_i = 0$ for complex crystal structures like Mn and $\varphi_i = -1$, if bcc structure.

The attrition milled powder was analysed by XRD analysis to identify the phases that may have formed after the attrition milling process. The XRD graph of the attrition milled powder is shown in Fig. 2 and no phase formation was observed during milling process.

The milled powder was sintered using HPS with permutations of pressure and time keeping temperature constant at 1100 °C. Figure 3 shows the 3D Scatter plot of density with respect to pressure and time (at 1100 °C). It was concluded that the

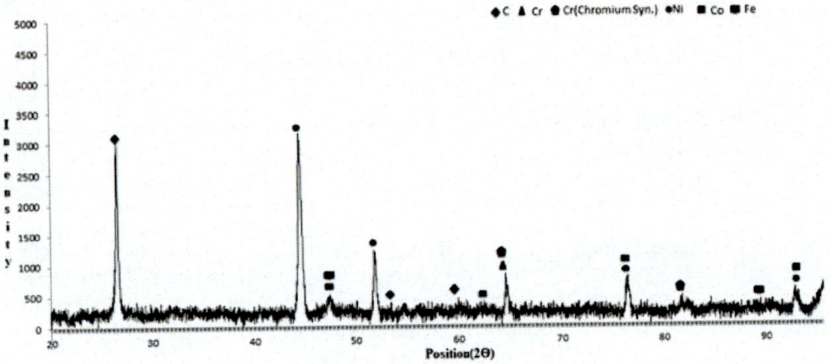

Fig. 2 XRD analysis of the milled elemental powder: $CoCrFeNiC_{0.2}$

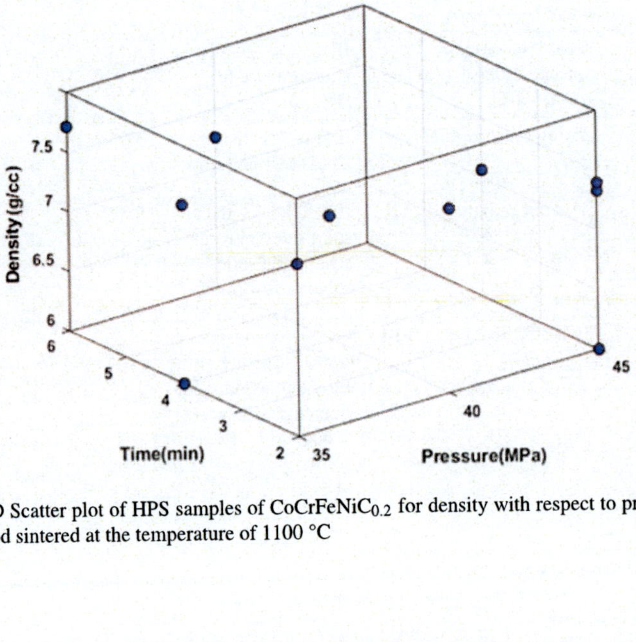

Fig. 3 3D Scatter plot of HPS samples of $CoCrFeNiC_{0.2}$ for density with respect to pressure and time period sintered at the temperature of 1100 °C

highest density of 7.71 gcm^{-3}, is obtained for the parameters 1100 °C, 35 MPa, 6 min.

The XRD graph of high density sample is represented in Fig. 4. It showed the presence of iron-nickel, carbon (bort and graphite), iron, nickel and cobalt-iron phases. It indicated that milled elemental powder has simple cubic, face-centred cubic and body-centred cubic and hexagonal closed packed solid solution phases. The phases cobalt iron (CoFe) and bort carbon have a simple cubic structure and iron-nickel (FeNi) and nickel have a fcc crystal structure and iron has a bcc crystal structure and and graphite carbon have a hcp crystal structure. Table 3 tabulates empirical formula, crystal structures, space group, lattice parameters, volume of unit cell and density of the phase. As given in Table 3 and from the XRD pattern (Fig. 4), iron present in the alloy is the solvent in the solid solution with atleast one of the elements—carbon, iron, nickel and chromium as the solute. Also, nickel present in the alloy is the solvent in the solid solution with atleast one of the elements—carbon, iron and cobalt as the solute. Further, the iron-nickel phase (FeNi) is also a solid solution with nickel as the solvent and iron as the solute. Also, the cobalt-iron phase (CoFe) is also a solid solution with cobalt as the solvent and iron as the solute. This has been stated on the basis of similar space groups in each of these solvent–solute pairs for a particular solid solution. All these solid solution possibilities have been made in accordance with the Hume-Rothery rules. Further, the carbon present in the alloy is in two forms—graphite and bort.

The XRD graph of HPS and then sintered (1300 °C for 4 h) sample in Fig. 5 shows the presence of cobalt, carbon, Nickel–Chromium-Iron, iron-nickel and FeC_8 phases.

The hot pressed samples were etched using Murakami reagent and observed under optical microscopy for various phase formation analysis. The phases in the XRD Analysis are in accordance with the SEM–EDS Analysis results and thermodynamic reactions. The SEM–EDS Analysis gives the spatial distribution of the elements in the alloy (Fig. 6).

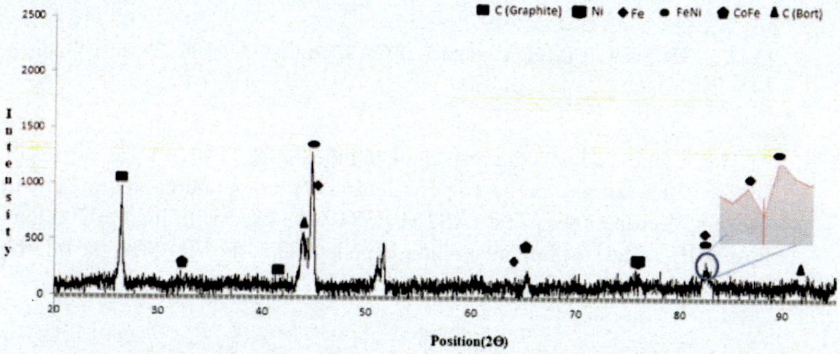

Fig. 4 XRD Analysis for hot press sintered samples of $CoCrFeNiC_{0.2}$ (1100 °C, 35 MPa, 6 min)

Table 3 Phases in hot press sintered pellet ($CoCrFeNiC_2$) (1100 °C, 35 MPa, 6 min)

Phase	Empirical formula	Crystal structure	Space group	Pearson symbol	Lattice parameters (Å)	Volume of unit cell (cm^3)	Density of the phase (g cm^{-3})
Cobalt-Iron	CoFe	Simple cubic	Pm-3 m	cP4	a = b = c = 2.857	23.32	8.27
Iron	Fe	Cubic; bcc	Im-3 m	cI2	a = b = c = 2.866	23.54	7.784
Nickel	Ni	Cubic; fcc	Fm-3 m	cF4	a = b = c = 3.557	45	8.9
Iron-nickel	FeNi	Cubic; fcc	Fm-3 m	cF4	a = b = c = 3.557	45	8.9
Carbon (graphite)	C	Hexagonal closed packed	P63/mmc	hP4	a = b = 2.456 c = 6.696	34.978	2.28
Carbon (bort)	C	Simple cubic	Fd-3 m	cF8	a = b = c = 3.567	45.37	3.52

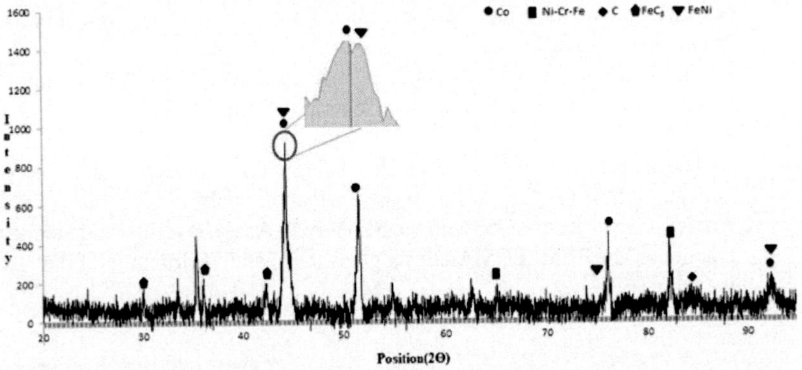

Fig. 5 XRD Analysis for HPS samples of $CoCrFeNiC_{0.2}$ (1100 °C, 35 MPa, 6 min) sintered at 1300 °C for 4 h

The hardness depicts the resistance to indentation. The test was carried out on the sintered HPS samples prepared permutations of temperature, pressure and time. The Vickers Microhardness Test (ASTM E384) was used with load = 300 g and dwell time = 10 s to test the hot press sintered pellets ($CoCrFeNiC_2$) and the microhardness values are given in Table 4.

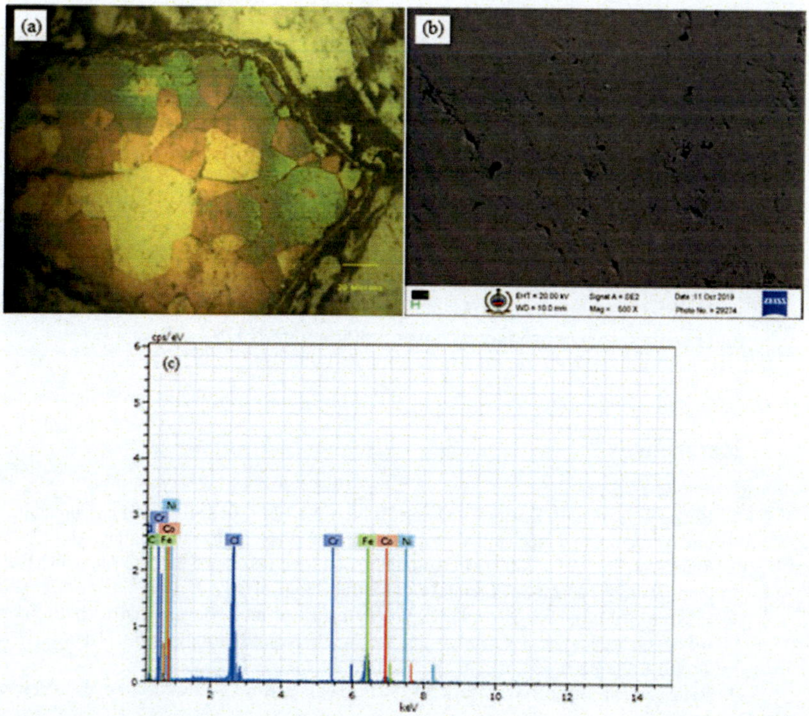

Fig. 6 Microstructure of sintered HPS sample ($CoCrFeNiC_{0.2}$)-sintered at 1300 °C, 4 h in Ar-atmosphere **a** Optical microscopy at 500 X- after etching **b** SEM image **c** SEM–EDS analysis

Table 4 Vickers microhardness test of the HPS and sintered samples ($CoCrFeNiC_{0.2}$) with permutations of temperature, pressure and time

Hot press sintering					Sintering heat treatment			Hardness
Weight of sample (g)	Diameter of sample, mm	Temperature (°C)	Pressure (MPa)	Time (min)	Temperature (°C)	Time (h)	Atmosphere	After
1.5	10	1100	35	6	1300	4	Argon	800–900 HV
1.6	10	1100	35	4	1300	4	Argon	500–610 HV
1.5	10	1100	35	2	1300	4	Argon	300–400 HV
2.5	10	1100	45	2	1300	4	Argon	55–100 HV
1.5	10	1100	45	2	1300	4	Argon	100–175 HV

4 Conclusion

High entropy alloys, themselves have excellent properties. But an additive can enhance the properties furthermore. Thus, certain additives are added to high entropy alloys to achieve optimum hardness. The analysis of the composition $CoCrFeNiC_{0.2}$ using optical microscopy, XRD and SEM suggest that the alloy contains some intermetallic compounds such as complex chromium carbides, etc., as well as a binary or a tertiary solid solution in the alloy. The presence of carbides is confirmed by the etchants (Murakami's reagent) used. The composition $CoCrFeNiC_{0.2}$ before sintering with HPS parameters 1100 °C, 35 MPa, 6 min has the highest density of 7.71 g cm^{-3} and has the highest hardness between 800 and 900 HV.

References

1. Huang TD, Jiang L, Zhang CL, Jiang H, Lu YP, Li TJ (2018) Effect of carbon addition on the microstructure and mechanical properties of CoCrFeNi high entropy alloy. 61:117–123
2. Shun T.-T, Du, Y.-C (2009) Microstructure and wear resistance of FCC $Al_{0.3}CoCrFeNiC_x$ high entropy alloys. Mater Sci Technol Conf Exhib 2154–2160
3. Praveen S, Murty BS, Kottada RS (2014) Effect of molybdenum and niobium on the phase formation and hardness of nanocrystalline CoCrFeNi high entropy alloys. J Nanosci Nanotechnol 14(10):8106–8109
4. Ranganathan S, Yeh J-W, Murty BS (2014) High-Entropy Alloys. Butterworth-Heinemann, pp 1–156
5. Chikumba S, Rao VV (2015) HEAs: development and applications. In: 7th International conference on latest trends in engineering and technology (ICLTET'2015). Irene, Pretoria (South Africa)
6. Kube SA, Sohn S, Uhl D, Datye A, Mehta A, Schroers J (2019) Phase selection motifs in high entropy alloys revealed through combinatorial methods: large atomic size difference favors BCC over FCC. Acta Mater 166:677–686

Modeling of Custom Patient-Specific Implants of Different Knee Joints Components Considering Different Materials

Amitesh Shrivastava, N. K. Jain, and R. Salhotra

1 Introduction

The human knee joint is hefty and regarded as one of the complex joints present in the human body. It contributes stability and meanwhile allows the movement of body. Knee joint is a modified hinge joint which causes flexion, extension and rotation by relative movement of femorotibial and patellofemoral joint [1]. The complexity of knee joint is such that it involves 3 large bones and one sesamoid bone including various ligaments, cartilage, meniscus and muscle groups. The bones that undergo formation of knee joint are femur, tibia, fibula and patella. Along with bones knee joint also consists of ligaments and tendons. Therefore it is regarded as largest joint in human body. Femur and tibia are attached by condyloid joint. The significance of tibiofemoral joint is to transmit the body weight under static and dynamic conditions from femur to tibia and means while it also provides rotation to the joint. Similarly, joint formed by patella and femur bones is patellofemoral joint. As compared to femur, patella bone is quite smaller and is enclosed in the patellar tendon. The posterior surface of patella is in contact with the anterior surface of femur condyles. While movement of knee joint relative motion is caused and friction between them is decreased by articulate cartilage. Moreover, in between femoral condyle and tibial plateau in knee joint a fibrocartilage, meniscus is present. [2]. The menisci act as a cushion and perform better shock absorption and load distribution in the knee joint. The menisci are thick on the outer side and thin on the inner side. Size of the menisci varies from person to person and it is observed that lateral menisci are thicker than medial menisci [3]. The shape of lateral menisci is circular, whereas the medial menisci are of crescent shaped [4]. The ligaments in knee joint keep bones aligned. The ligaments are attached such that they restrict the movement of bones and hold them in proper position. Important ligaments of the knee joint are collateral (lateral

A. Shrivastava (✉) · N. K. Jain · R. Salhotra
Mechanical Engineering Department NIT, Raipur 492001, India

R. Prasad et al. (eds.), *Advancement in Materials Processing Technology*,
Springer Proceedings in Materials 12,
https://doi.org/10.1007/978-981-16-3297-6_23

and medial) and cruciate ligaments (anterior and posterior). The lateral and medial collateral ligament restricts movement in side to side direction. One the other hand the cruciate ligaments (PCL and ACL) is liable for restricting sliding of femur on tibial plateaus while extension and flexion. Moreover, there are also the ligaments and muscle groups that are responsible for restricting the dislocations and injury to certain extent.

One of the most frequent injuries at the knee joint is meniscal tears and trauma (Fan and Ryu, 2000). The reasons for this may be due to accident, sports injury or degeneration due to ageing. Despite bones and meniscus can repair itself along the time but due to lack of blood supply and nutrition it does not heal properly [5]. In order to restore appropriate function, implant may be helpful which will reduce the pain and tear at the knee joint [6].

Human joints are degenerative in nature. As person ages the muscles or soft tissue starts to get stiff which cause pain and trauma [7–9]. In order to get relief surgery is performed after examination of the conditioning of the joint. Even after surgery, it is observed that it take long time to restore range of motion and proper gait. Implants available in Indian markets are generally imported which is quite bigger in size because it targeted the local population because of this even after playing it inside human body it take long time for rehabilitation.

2 Methodology

2.1 Computed Tomography (CT) and Magnetic Resonance Imaging (MRI) Scanning

Computed Tomography (CT) scanner machine emits the rays onto the area under examination from multiple angles. Computed Tomography (CT) scanning is a painless and non-incision practice which sequences multiple 2D X-ray images from different angles. CT scan generally develops a detailed image of the hard tissues undergoing investigation. Whereas, MRI-scans are preferred to observe a person's soft tissues. MRI helps in getting a detailed view of ligaments, menisci and cartilages. In current study, the healthy person undergoing CT scan is of age 28 years with no earlier physiological condition. 512 images were collected from the scan with slice thickness of 1.5 mm. Further, these images were utilized for creating realistic 3D model of the bones involved in knee joint. Similarly for modeling the soft tissues in knee joint MRI scan is performed, just like earlier 3D model of meniscus and cartilages were constructed using software. This method of developing 3D model is highly accurate and involves greater extent of diagnosis (Figs. 1 and 2).

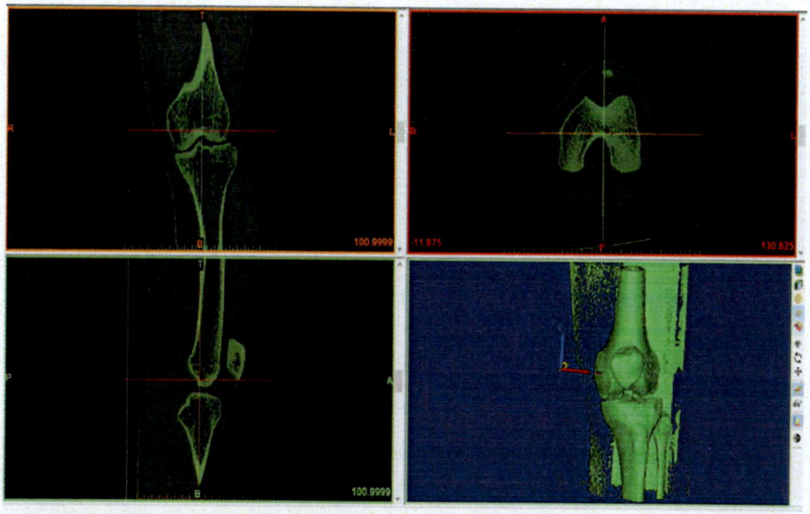

Fig. 1 3D model of knee joint constructed from DICOM files of CT Scan

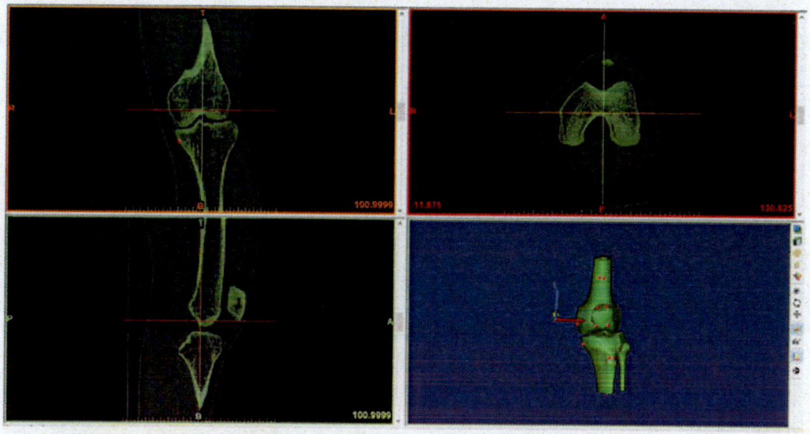

Fig. 2 3D model after reducing noise and eliminating the undesired pixels

2.2 Conversion of DICOM Data into 3D Model

The materialize mimics and 3-matics are one of the leading and accredited software for developing and medical models. These software are quite expensive and require some skill in order to perform operation. All the collected DICOM files are arranged and imported to materialize mimics software package for constructing the real scale 3D model of the bone, meniscus and cartilages structures. The noisy 3D model is

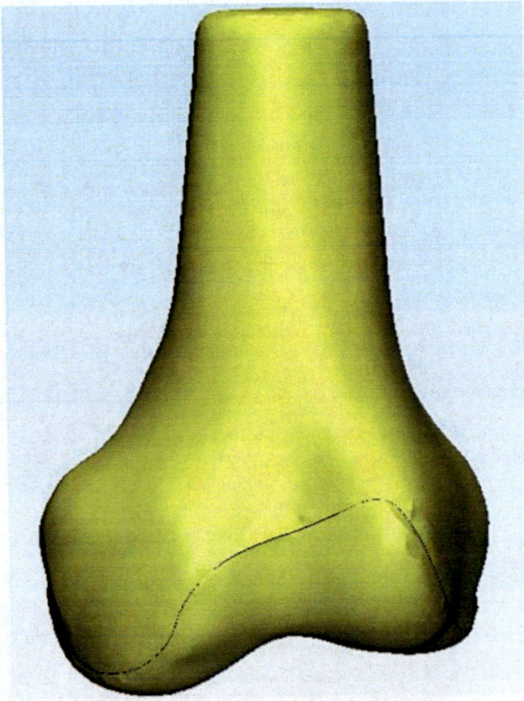

Fig. 3 3D model of femur head

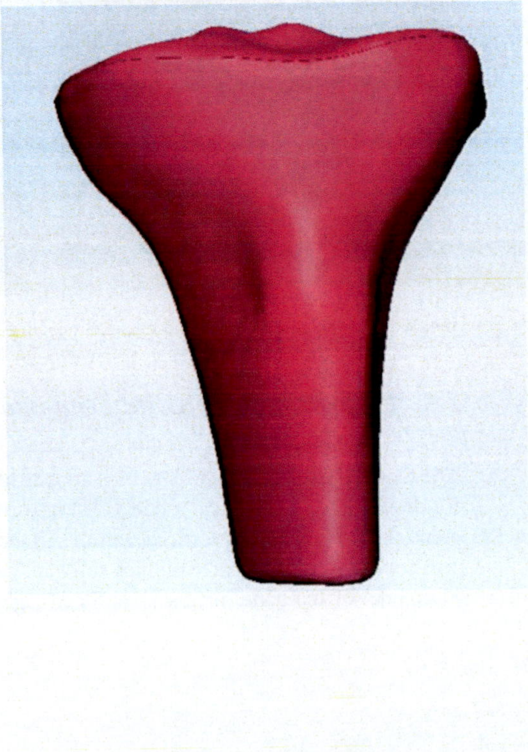

Fig. 4 3D model of Tibial head

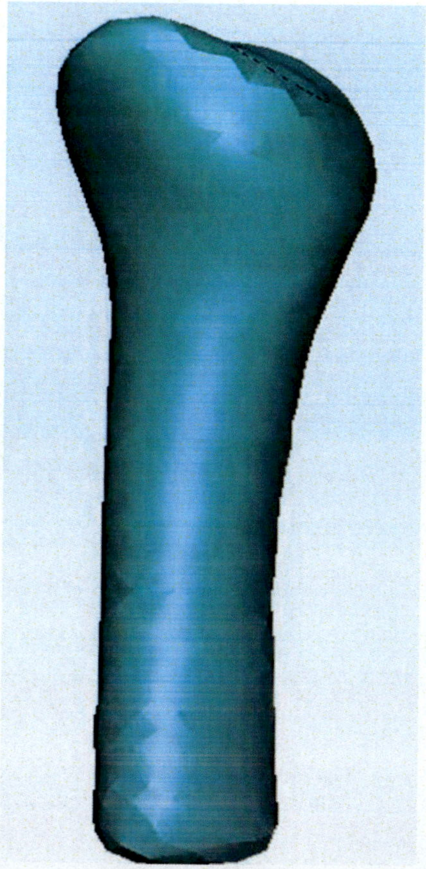

Fig. 5 3D model of fibular head

then segmented and separated from each other for construction of better and accurate geometry. After which the mask is created and pixels are arranged, the time for performing this operation may vary as per the operator's skill. From materialize mimics the models are further transferred and processed in 3-matic software package (Figs. 3, 4, 5, 6, 7, 8, and 9).

2.3 *Transformation of Raw 3D Model into STL Model*

The raw 3D model developed from the CT/MRI files is quite noisy and discrete at some point which requires further post-processing in order to develop an efficient structure. Multiple operations are performed in order to make geometry smooth. Holes are filled and sharp edges are smoothened in post-processing.

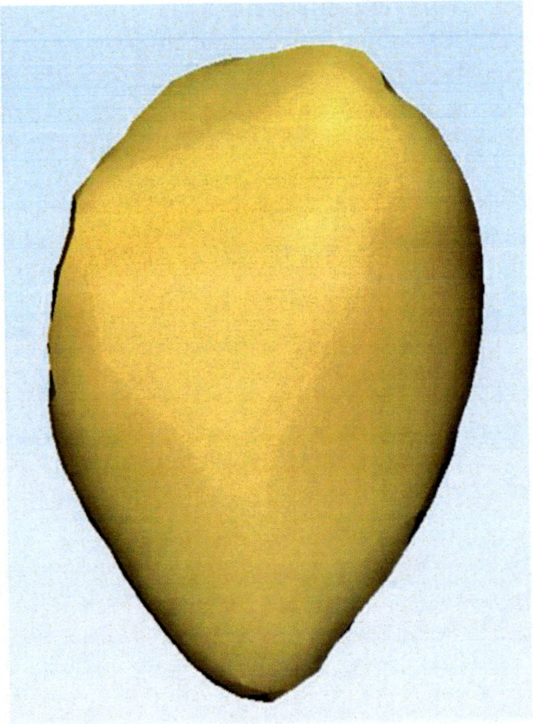

Fig. 6 3D model of patella

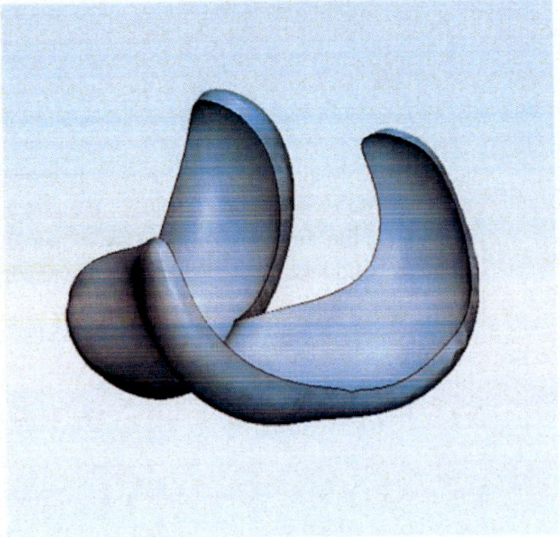

Fig. 7 3D model of femoral cartilage

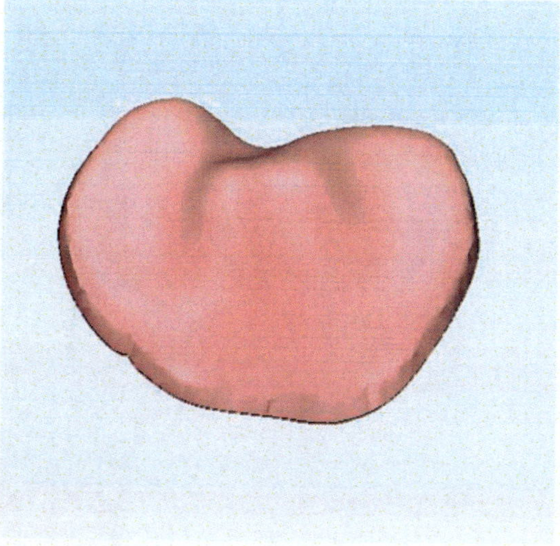

Fig. 8 3D model of tibial cartilage

2.4 3D Printer and FFF Fabrication

The additive manufacturing system is recent technique which can be utilized for creating accurate custom implants for human body [9]. There are two basic steps involved in developing custom implants: (1) creation of 3D medical bio-model of implant and (2) FFF system for design and manufacturing of implant.

The 3D printer used for manufacturing the product is extrusion type and Polylactic Acid (PLA) material is used because of its biocompatibility. Widely the implants or prosthetic parts for human body are manufactured from Ti6Al4V alloy. The Ti6Al4V implants are much costlier and it is very difficult to manufacture for specific patients. Therefore PLA customized implants are developed which is a lot cheaper and quite efficient to perform well.

3 Observations and Results

3.1 Development of Knee Joint-Specific Implant

Additive manufacturing or Rapid prototyping is advance manufacturing technique which allows us to manufacture accurate physical models using CAD software and 3D printer. While development phase cost of manufacturing can be varied depending on different parameters, i.e., layer thickness, orientation, supports and material, etc. Thus for fabricating model different critical parameters are optimized also the surface

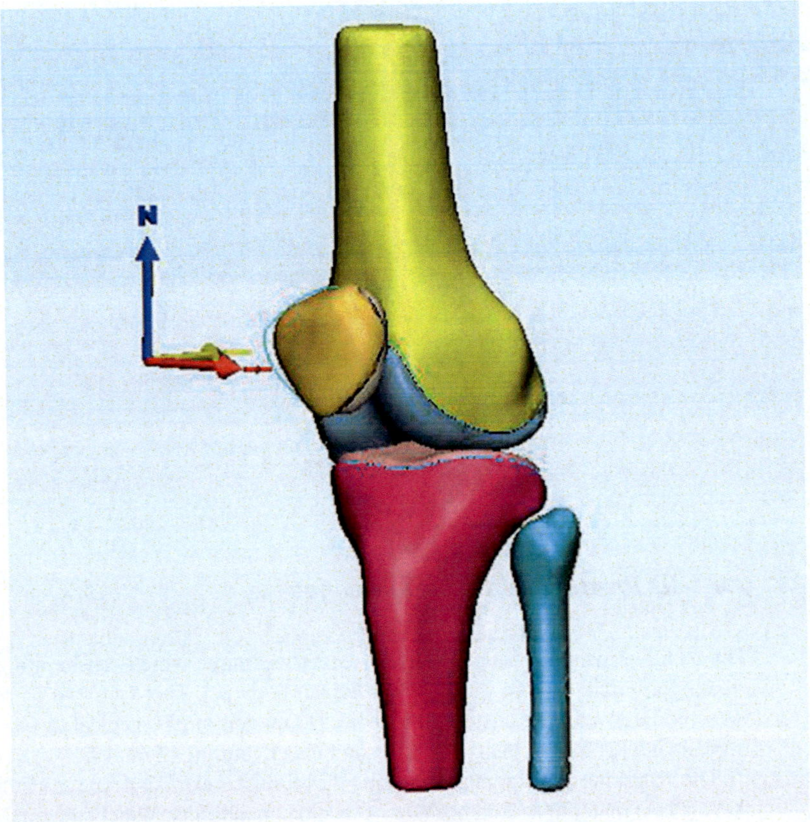

Fig. 9 3D model of assembled knee joint

finish and accuracy depend on these parameters. The PLA implants may be directly used inside the body after sterilizing and placing in accurate position through surgery.

3.2 *Numerical Analysis*

Besides just developing the implants numerous aspects of the bio-structures or implants can be analysed numerically. It is obvious that before mass production or using an implant for specific operation a numerical analysis is performed in order to justify the quality of the product. As the organic surfaces are irregular in shape and they have surface discontinuity it is very much important to select specific mesh that can simplify its surface. For meshing tetrahedron- 10 meshes are used for meshing (Fig. 10).

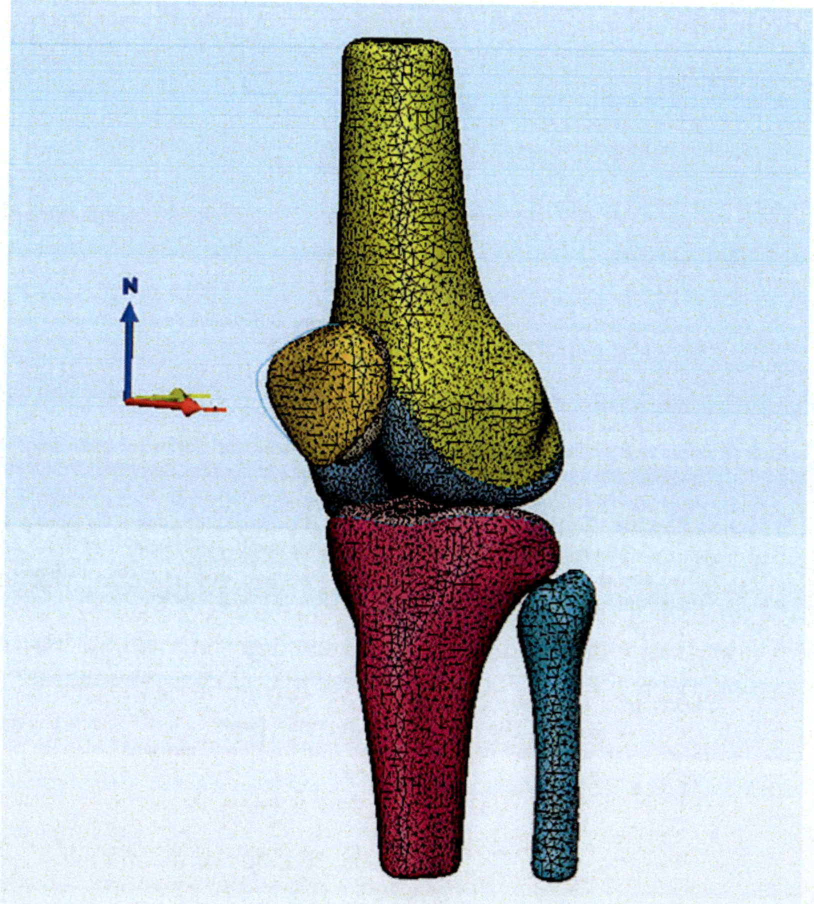

Fig. 10 Meshed 3d model of knee joint

4 Discussions

From current paper following conclusions can be carried out:

1. The effective and accurate development of 3D model from the DICOM files is accomplished.
2. Post-processing of the generated 3D model helps in getting more realistic and better surface finished product.
3. The developed PLA 3D model of implant is biocompatible and can be placed inside through surgery.
4. Present study can also be used further for carrying out different numerical analysis.

Acknowledgements No funds were received in support to this study. No benefit in any form has been or will be received from a commercial party related directly or indirectly to the subject of this manuscript

Conflict of Interest There are no conflict of interest related to the manuscript.

References

1. Goldblatt JP, Richmond JC (2003) Anatomy and biomechanics of the knee. Oper Technol Sports Med 11(3):172–186
2. Hall SJ (2011) Basic biomechanics, 6th edn. McGraw-Hill, New York
3. Makis EA, Hadidi P et al (2011) The knee meniscus: structure function, pathophysiology, current repair. Biomaterials 32(30):7411–7431
4. Brindle T, Nyland J, Johnson DL (2001) The meniscus: review of basic principles with application to surgery and rehabilitation J Orthop Trauma 30(3):232–238
5. Athanasiou KA, Sanchez-Adams J (2009) Engineering the knee meniscus. Morgan & Claypool
6. Galley NK, Gleghorn JP, Rodeo S, et al (2011) Frictional properties of the meniscus improve after scaffold-augmented repair of partial meniscectomy: a pilot study
7. Navacchia A, Humea DR, Rullkoetter PJ, Shelburne KB (2019) A computationally efficient strategy to estimate muscle forces in a finite element musculoskeletal model of the lower limb. J Biomech
8. Gokkus K, Atmaca H, Uğur L, Özkan A, Aydin AT (2016) The relationship between medial meniscal subluxation and stress distribution pattern of the knee joint: finite element analysis. J Orthop Sci 27:32–37
9. Lu Y, Li J, Miller SC, Jin Z, Hua X (2019) Development of a finite element musculoskeletal model with the ability to predict contractions of three-dimensional muscles. J Biomech 94(20):230–234
10. Balwan AR, Shinde VD (2019) Development of patient specific knee joint implant. Mater Today Proc
11. Lian Q, Li D (2014) Patient-specific design and biomechanical evaluation of a novel bipolar femoral hemi-knee prosthesis
12. Guess TM, Thiagarajan G, et al (2010) A subject specific multibody model of the knee with menisci. Med Eng Phy 32:505–515
13. MeenakshiK CVR, Kumar KS, Ramana SV (2020) Tribological aspects on human knee joint—a review. Mater Today Proc 22(Pt 4):3100–3105
14. Khoshgoftar M, Vrancken ACT, van Tienen TG, Buma P, Janssen D, Verdonschot N (2015) The sensitivity of cartilage contact pressures in the knee joint to the size and shape of an anatomically shaped meniscal implant. J Biomech 48(8), 1427–1435
15. Namin AT, Jalali MS, et al (2029) Adoption of new medical technologies: the case of customized individually made knee implants. Value Health 22(4):423–430
16. Zapata G, Sanz-Pena I, et al (2019) Effects of femoral component placement on the balancing of a total knee at surgery. J Biomech 86:117–124
17. Zapata G, Morton J (2020) Principles of a 3D printed mechanical device for total knee balancing. J Biomech 112:110039

Boiling and Condensation Heat Transfer Assisted Separation Phenomena During Distillation Refining of Mg Metal Using Thermodynamics and Numerical Simulations

Krishna Kumar, Suchandan K. Das, Snehashish Tripathy, S. R. M. Prasaana, and Manoj Kumar

1 Introduction

Vacuum distillation technology is being explored currently, as metal distillation is an interesting physical separation technique to obtain high pure metals. During distillation, higher vapour pressure elements or impurities can be distilled off for separation. Refining by metal–gas processes is greatly facilitated by the use of vacuum [1, 2]. The Clausius–Clapeyron equation establishes the importance of vacuum in refining through vacuum distillation process [3, 4]. The use of vacuum makes it possible to carry out refining processes at lower temperatures. Metals having boiling point below 1000 °C that may be refined by distillation [1, 2]. Above this temperature, practical operational problems increase considerably. Higher melting point metals can be refined, using vacuum distillation method. Metals with a vapor pressure approximating to that of pressure under vacuum can be continuously be distilled off. The selective distillation of metals in different thermal zones can be useful to separate two or more metals from a solid solution/mixture.

K. Kumar (✉) · S. K. Das · S. Tripathy · M. Kumar
CSIR-National Metallurgical Laboratory, Jamshedpur, Jharkhand 831007, India
e-mail: kkrishna@nmlindia.org

K. Kumar · S. K. Das · M. Kumar
Academy of Scientific and Innovative Research (AcSIR), CSIR-HRDC, Ghaziabbad, UP 201002, India

S. R. M. Prasaana
Metallurgical Engineering Dept., Government College of Engineering, Salem, Tamilnadu 636011, India

R. Prasad et al. (eds.), *Advancement in Materials Processing Technology*, Springer Proceedings in Materials 12,
https://doi.org/10.1007/978-981-16-3297-6_24

2 Literature

From basic thermodynamics, the Clausius–Clapeyron equation establishes the relationship of temperature and vapour pressure of metals, which has been explored by many researchers in various fields of research. Drápala and Kuchar [5, 2] have reported an approximate classification of the Metals based on the vapour pressure of their saturated vapours. They have shown that it is possible to evaluate the potential capacity for separation of these metals by distillation. Vapour pressure differences and the resulting relative volatility describe the separation feasibility of impurities or components from a mixture. Refining of these metals by subsequent rectification distillation has also been technically established [1, 2]. Vacuum distillation has been used for removing the impurities, from industrially refined magnesium [6]. Distillation results in lowering of the concentration of every impurity in the useful part of the distillates. Impurities more volatile than magnesium and less volatile than magnesium behave differently due to significant relative volatility (α) among both purity and impurity. In the experiments carried out by [6], four factors, i.e., load, temperature, time, and purity have been investigated to establish the correlation among them. Wang etal. Has presented a thermodynamic analysis and estimated the degree of the separation between magnesium and impurities under vacuum with Clausius–Clapeyron equation [7, 8]. They had demonstrated the feasibility of vacuum distillation on experimental scale for Mg refining. They were able to remove the low-volatile impurities such as Fe, Cu, Mn, Pb, Ca at the stage of 600 ^{0}C for 30 min. The feed magnesium was refined from 96.0% (1N6) feed magnesium to that of 99.98% (3N8) purity and total impurity content was reduced from ~34,600 to ~180 ppm upon condensate [8]. Researchers have also studied the production of high purity magnesium of 99.99% by vacuum distillation using commercial grade magnesium and magnesium alloy AM100 at laboratory scale. Researchers have studied this technique for a range of metals at lab scale.

The potential and uniqueness of distillation refining and the available expertise on the development of Electrothermal process for magnesium metal production at pilot scale of 450 kg raw material (40 kg Mg/batch), leads us to investigate the thermodynamics and heat transfer associated in boiling and condensation of metal vapours inside electrothermal condenser. In view of scarcity of knowledge available for boiling and condensation of metals, this study develops an understanding of separation using selective boiling and heat transfer controlled condensation towards high purity metals. The understanding would be helpful in devising the boiling and condensation based distillation strategy for low boiling point-metal like Mg metal refining. The parameters studied may help to choose conditions as well as materials for controlled-condensation w.r.t heat transfer during phase change w.r.t dropwise condensation aspects.

2.1 Thermodynamic Analysis for Boiling

Distillation is carried out on the basic principle of relative volatilities of the components of a mixture at specific pressure and temperature conditions. The vapour pressure as well the vaporization and condensation behaviour controls the overall distillation process. For evaporation at equilibrium, among liquid (l) and gaseous (g) phases, the equation can be expressed as following equations [1, 9, 2]. Assuming, activity of pure liquid metal = 1, where $p_m{}^0$ is partial pressure of pure metal (M) at Temperature (T),

$$ln\ \ p_m{}^o = -\Delta G^0/RT \tag{1}$$

$$ln\ \ p_m{}^o = -(\Delta H^0 + T\Delta S^0)/RT \tag{2}$$

Since, ΔS^0 of evaporation of metals is nearly constant and ΔH^0 of evaporation or latent heat of evaporation (Le) varies orderly across periodic table,

$$ln\ \ p_m{}^o = -\frac{L_e}{RT} + C \tag{3}$$

At standard state conditions, the variation in the vapor pressure of a pure metal against temperature is usually approximated by Clausius–Clayperon equation [1, 2, 4]. The mathematical expression of the dependence of the pressure of saturated the vapor (p) of a substance on temperature (T) is given in Eq. (4). The vapour pressure (p) of metal increases with increase in temperature of a molten metal [1, 5, 10, 2, 11].

$$\ln p = \frac{\Delta H_{ev(0)}}{RT} + \frac{\alpha}{R}\ln T + \frac{\beta T}{R} + \ldots I \tag{4}$$

where $\Delta H_{ev(0)}$ is the molar enthalpy of evaporation of the substance at temperature T = 0 K; α, β are temperature coefficients; I is the integration constant. Further, Eq. (4) can be simplified to plot the relationship among P and T as Eq. (5). Equation (5) can be used to estimate vapour pressure of components of a Mg-impurity matrix. Figure 1 depicts the variations of vapour pressure against temperature for some low boiling metals.

$$\ln p = \frac{A}{T} + B \tag{5}$$

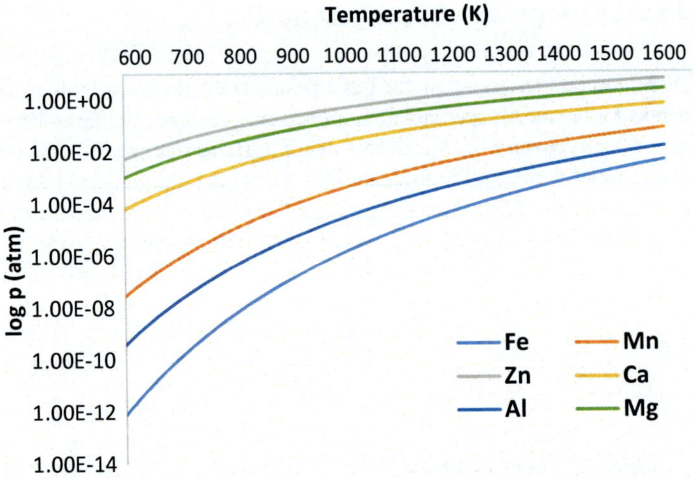

Fig. 1 Variation in vapour pressure (log p) of elements with temperature (548–1548 K) at 1 atm (ref: Alcock, CB; Itkin, VP and Horigoan MK, Canadian Metallurgical Quarterly, 23 309 1984)

2.2 *Relationship of Activity and Vapour Pressure of Components in Mixtures*

The activities of metal in alloys, like activities of individual components in the solutions, determine the intensity of the boiling as well as condensation processes. The vapor pressure of a component in an alloy P_i is related to the vapor pressure of the pure system, i.e., metal P by the following Eq. 6. [2, 4]

$$P_i = a_i P_i^0 = \gamma_i X_i P_i^0 \tag{6}$$

where a_i is the activity of the metal in the alloy, γ_i is its activity coefficient, and X_i is its mole fraction. The vapor pressures any element in a system (solution/mixture), in their pure states may be of the same order of magnitude. For any solution, the vapour pressure of the impurity (over the solution) is controlled by the composition (Xi). If the impurity forms an ideal solution, then $\gamma_i = 1$, but even than P_i will be less than P because X_i is less than unity. When the impurity shows a positive deviation from the ideal behaviour, $\gamma_i > 1$ and the decrease in vapor pressure over the solution due to X_i being less than unity is partly offset. Thus, based on vapour pressure criteria, the distillation can be applied for separation. Fractional distillation may be used, if the vapour pressure difference is small for good purification [5, 10, 2, 4]. One of the important factors that determine the efficiency of separation is the relative volatility of the components expressed in terms of their partial pressures p_A and p_B of A and B, respectively over solution, then considering Eq. 8 for A and B,

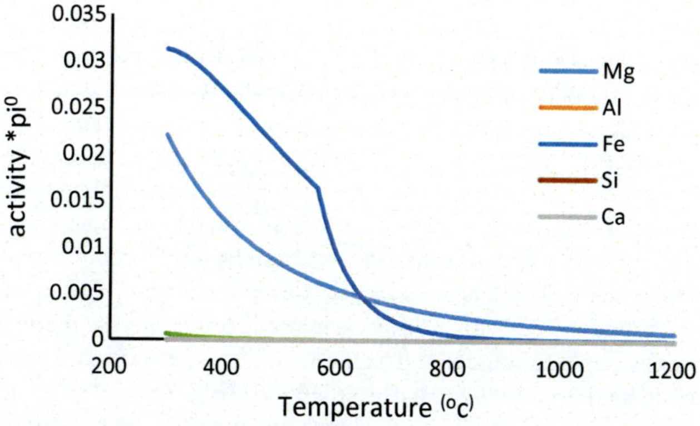

Fig. 2 Variation in individual activity of individual components with temperature

$$\frac{p_A}{p_B} = \frac{\gamma_A X_A P_A^0}{\gamma_B X_B P_B^0} = \frac{a_A P_A^0}{a_B P_B^0} = \alpha \tag{7}$$

where, α is the relative volatility. It is evident that α is a direct measure of the ease with which the components can be separated by means of distillation. Temperature changes have an almost similar influence on the vapor pressures of all the components in the system, α usually does not vary. It only varies with activity coefficients of components. ThermoCalc software was used to calculate the activity of constituents based on 99% Mg and rest impurities (Ca, Si, Al, Fe) at standard state. Further Eq. (6 and 7) was being used to estimate the partial pressures and changes thereof. The obtained data was used to estimate the relative volatilities in the system w.r.t Mg metal (Fig. 2).

2.3 *Condensation Heat Transfer*

In electrothermal process, the saturated magnesium vapours produced in reactor enter the condenser at saturation temperature 1573 K by virtue of process [12, 13, 14]. Thermal gradients are the significant driving force for heat transfer from T_{sat} (condenser inner surface) to T_w (condenser outer body). The Electrothermal process condenser has been chosen as the system of condensation, which is basically a cyclindrical tube in vertical alignment [15, 13]. Dropwise condensation (DWC) transient heat transfer model for single droplet growth has been numerically investigated for Mg-impurity vapour–liquid system in condenser.

Since, DWC facilitates higher heat transfer rate than that of a FWC [16, 13, 17]. The heterogeneous DWC condensation phenomena can be studied for selective condensation of metal vapours from a given mixture to get high purity metals, which has been studied in present case. The assumptions have been incorporated in the heterogeneous dropwise condensation model, which has been explained elsewhere [16, 13, 18]. With suitable assumptions, a heat transfer model had been developed for a single metal droplet system in vapour–liquid environment without consideration nucleation, detachment or coalescence phenomena. An equivalent spherical-cap approximation has been incorporated to droplet shape in the model. Thermodynamically constrained smallest radius is assumed as the minimum radius in the simulation [16, 19]. Mode of heat transfer is assumed to be purely conduction, "under saturated vapour system". Thermal resistances arise due to the liquid–vapor interface, curvature and conduction in substrate, different material properties and thermal inhomogeneity, driven by imposed sub-cooling on the substrate. The accommodation coefficient was taken to be 0.5 for liquid metals studied [16] (Fig. 3).

The assumed system is a one-dimensional steady-state conductive heat transfer. The geometry is considered as standard spherical shell (near pendent drop model [17] consisting of multiple layers offering thermal resistances (R_i). For heat transfer, the governing Fourier's heat conduction equation is given as Eq. 8 [9, 3]. Considering all significant thermal resistances, described elsewhere [13, 16, 20 and 21] the rate of heat transfer based growth rate of metal droplets contained in the system described above may be given as equation (Eq. 9).

$$q_d = -\frac{(T_{sat} - T_w)}{\sum R_i} \tag{8}$$

$$\frac{dr}{dt} = \left(\frac{4(T_{sat} - T_w)}{\rho_l h_{lv}}\right)$$

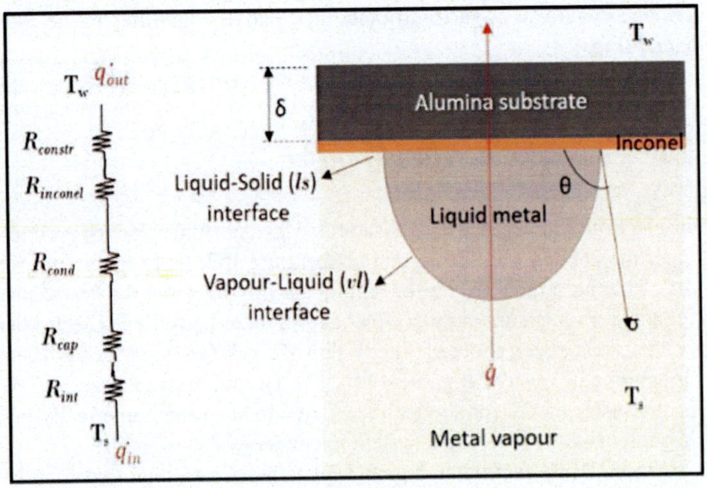

Fig. 3 Pendent droplet model of liquid metal with thermal resistances in series during condensation heat transfer [13, 18]

$$X\left[\frac{\left(1-\frac{r_{min}}{r}\right)}{\left(\frac{2}{h_{int}}\right)+\left\{\frac{r\left(1-cos\theta_{avg}\right)}{k}\right\}+\left\{\frac{4\delta}{k_{coat}\left(1+cos\theta_{avg}\right)}\right\}+\left\{R_{constr}.4\pi r^{2}\left(1-cos\theta_{avg}\right)\right\}}\right]$$

$$X\left[\frac{\left(1-cos\theta_{avg}\right)}{\left(2-3cos\theta+cos^{3}\theta\right)}\right] \quad (9)$$

The rate of growth of individual drop for horizontal surfaces with and without wettability gradient can be estimated using above ODE [13, 16 and 22]. The above equation (Eq. 9) incorporates the effect of constriction resistance on heat transfer and consequent the droplet growth during condensation [13].

3 Numerical Analysis of Condensation Heat Transfer

Using Clausius–Clapeyron equation and the separation coefficient parameter (relative volatility) the feasibility to separate the mix of metals is established. Thermo-Calc® software has been used to estimate the standard state activities of various metals. A computational Matlab® code has been developed to solve the governing Eq. 9 and ODE was solved for various parameters under consideration. A 4th order Runge–Kutta Predictor–Corrector method [23] has been used to solve the system of differential equations. The parametric sensitivity analysis was conducted. The droplet growth dynamics during heterogeneous condensation were estimated with respect to differential temperature (ΔT), contact angle (θ) with the surface, thickness of substrate (δ) and ratio of inactive to active surfaces area on the condenser surface (β) for all metal vapour–liquid systems. Table 1 depicts the thermo-physical properties of major impurities considered in study.

4 Results and Discussions

4.1 Thermodynamic Feasibility Through Separation Coefficient (A)

Thermodynamic calculations using ThermoCalc® software have been presented below in Fig. 4. Feed composition is taken as Mg ~98.9% and remaining Al, Ca, Si, and Fe. Activities of individual metals were estimated with ThermoCalc® activity coefficients (γ) database at standard state conditions. The relative volatility depicts that except Fe, all the elements relative volatility decreases w.r.t Mg. However, for Fe above 700–800 °C the relative volatility increases as the temperature dominates the

Table 1 Properties for various metals at liquid phase used in numerical analysis

Properties (Unit)	Silicon	Calcium	Aluminium	Iron
Thermal conductivity (W/mK) $	56	200	91	33
Melting point (K)	1683	1115	933.5	1811
Boiling point (K)	3538	1757	2743	3135
Molecular weight (Kg)	0.028085	0.040078	0.026982	0.055845
Latent heat (J/Kg)	1.28×10^7	3.87×10^6	1.09×10^7	6.26×10^6
Density £ (Kg/m^3)	$2.524-(3.487 \times 10^{-4} \times (T-T_m)) \times 1000$	$1.613-(2.21 \times 10^{-4} \times T) \times 1000$	$2.378-(10.51 \times 10^{-4} \times (T-T_m)) \times 1000$	$7.035-(9.26 \times 10^{-4} \times (T-T_m)) \times 1000$
Surface tension # (N/m)	$850-(0.25 \times (T-1683)) \times 0.001$	$472-(0.001 \times T) \times 0.001$	$1050-(0.25 \times (T-933.5)) \times 0.001$	$1880-(0.41 \times (T-1811)) \times 0.001$

Ref: *(B S [18] #24 $[28] £(Busey, n.d.)

Mg vaporization compared to Fe. This may be studied towards selection of temperature of boiling operation. With calculated selection, impurities vapourization can be controlled with respect to magnesium, which will yield more pure Mg in condenser during condensation process. It should also be noted that separation coefficient, i.e., relative volatility of each system is sufficiently high to qualify distillation as a tool of separation.

The relative volatility depicts that, except Fe, all the elements relative volatility decreases w.r.t Mg. However, for Fe above 700–800 °C the relative volatility increases as the temperature dominates the Mg vaporization compared to Fe. This may be studied towards selection of temperature of boiling operation. With calculated selection, impurities vapourization can be controlled with respect to magnesium, which will yield more pure Mg in condenser during condensation process. It should also be noted that separation coefficient, i.e., relative volatility of each system is sufficiently high to qualify distillation as a tool of separation.

4.2 *Effect of Differential Temperature ($\Delta T = T_S - Tw$) on Rate of Droplet Growth*

The effect of differential temperature has been numerically investigated for all major impurity metals Ca, Al, Fe, Si usually present in commercial grade Mg metal. By

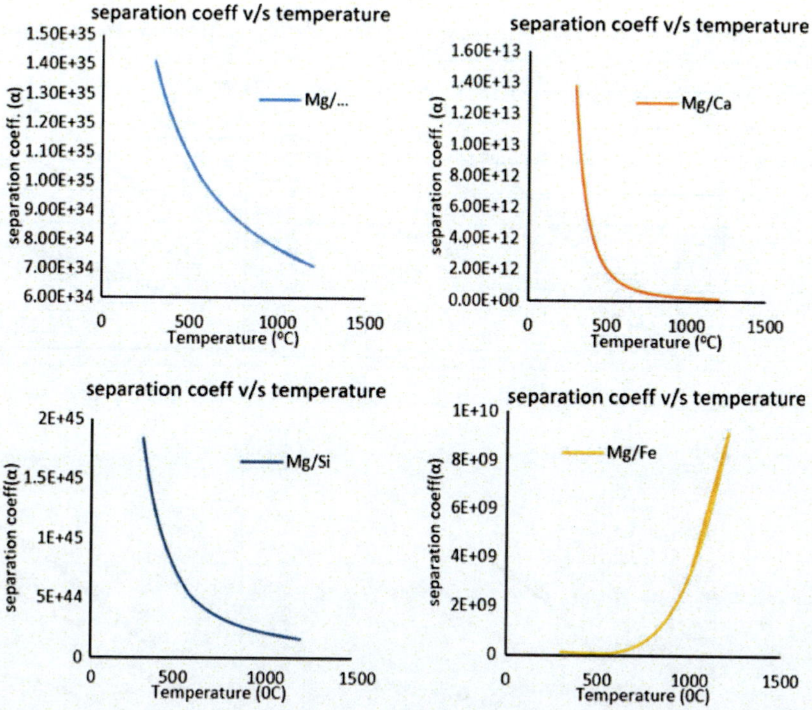

Fig. 4 Relative volatilities estimated for various Mg-impurity system calculated based on ThermoCalc software predictions and composition of system with Mg ~99%

variation of β, δ, and ΔT, the condensation behaviour was studied through DWC integrated model developed with considering the constriction resistance phenomena [13, 21].

Figure 5 above depicts that, the droplet growth dynamics for major metal impurities under varying thermal gradient conditions (ΔT). It has been observed that with higher driving force, i.e., thermal gradients (ΔT), the growth is higher. This may be explained as lowering in overall heat transfer from vapour to wall through liquid metal. Thermal gradient strongly controls the condensation phenomena, but relatively the condensation behaviour is unique and distinct for all metal impurities. Figure 5 dictates that qualitatively for the same time period growth of droplet growth in terms of size (m) will be of Al > Ca > Si > Fe order.

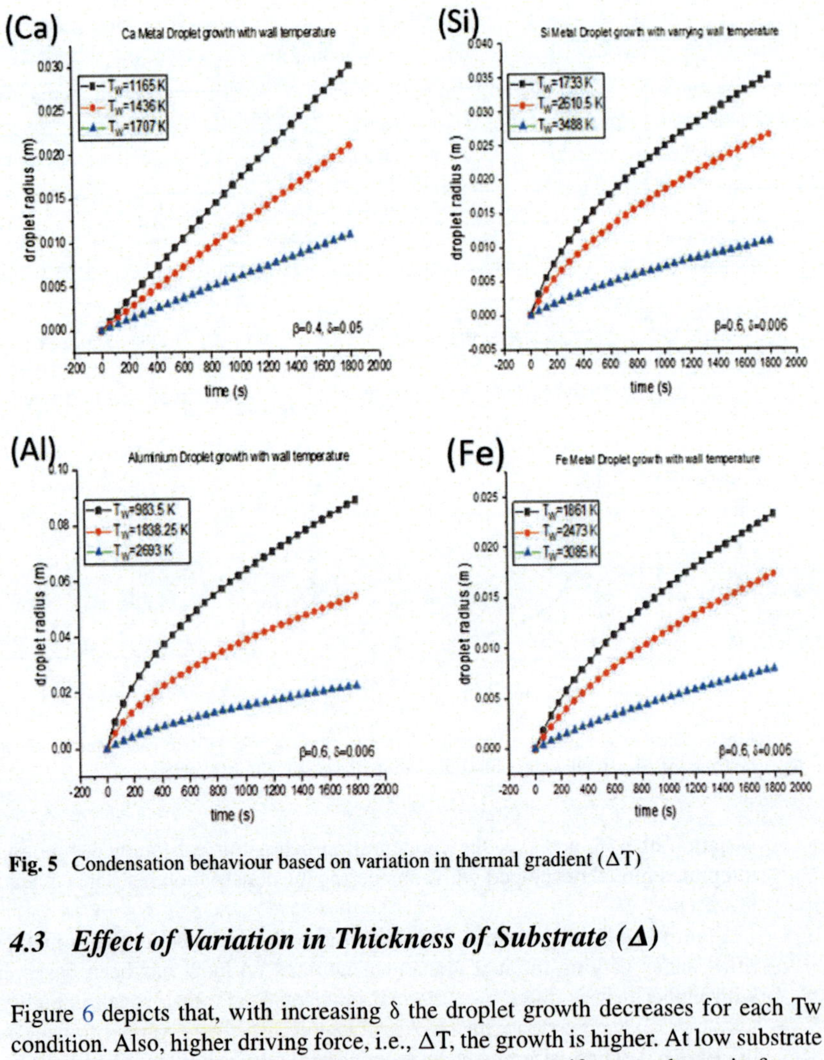

Fig. 5 Condensation behaviour based on variation in thermal gradient (ΔT)

4.3 *Effect of Variation in Thickness of Substrate (Δ)*

Figure 6 depicts that, with increasing δ the droplet growth decreases for each Tw condition. Also, higher driving force, i.e., ΔT, the growth is higher. At low substrate thickness, the thermal conductivity of substrate is not uniform as shown in fig., so the constriction resistance is small compared to larger substrate thickness, which accounts for high heat transfer rate at low substrate thickness compared to high substrate thickness. The heat flow lines become uniform in high substrate thickness plate and also the constriction resistance is large [21, 25]. The growth rate at $\delta = 0.005$ m is thrice greater than the growth rate at $\delta = 0.025$ m. This shows that constriction resistance varies gradually in slight variation in thickness of the substrate when the substrate thickness is small and it controls the condensation heat transfer significantly for metal condensation phenomena.

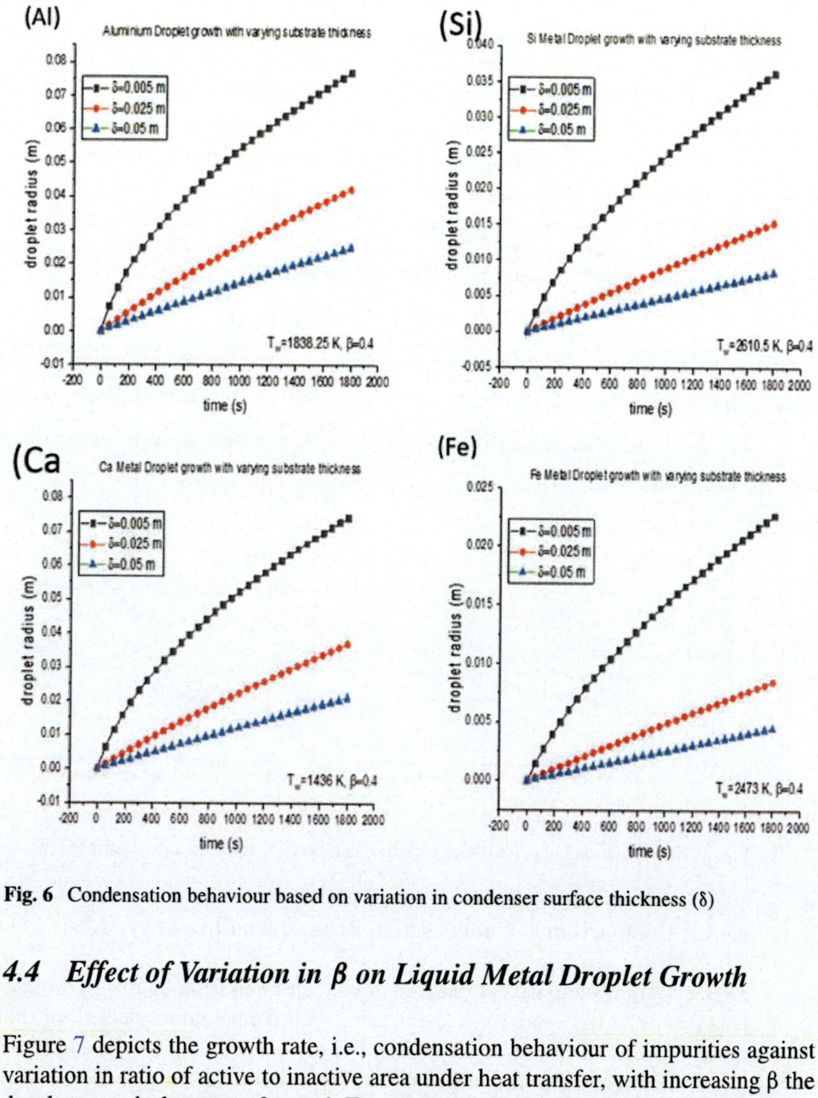

Fig. 6 Condensation behaviour based on variation in condenser surface thickness (δ)

4.4 *Effect of Variation in β on Liquid Metal Droplet Growth*

Figure 7 depicts the growth rate, i.e., condensation behaviour of impurities against variation in ratio of active to inactive area under heat transfer, with increasing β the droplet growth decreases for each Tw condition. In addition, higher driving force, i.e., ΔT, the growth is higher. At low β value, the thermal conductivity of the substrate is high because the heat transfer area and inactive area is small, consequently the constriction resistance is also small. At large β value, the inactive area is large and the thermal conductivity of the substrate is low. So the constriction resistance is large. At lower temperature gradient and lower β value, the growth rate becomes linear which states that the thermal conductivity and constriction resistance does not vary to a great or significant extent. At higher β value, the growth rate variation is much larger.

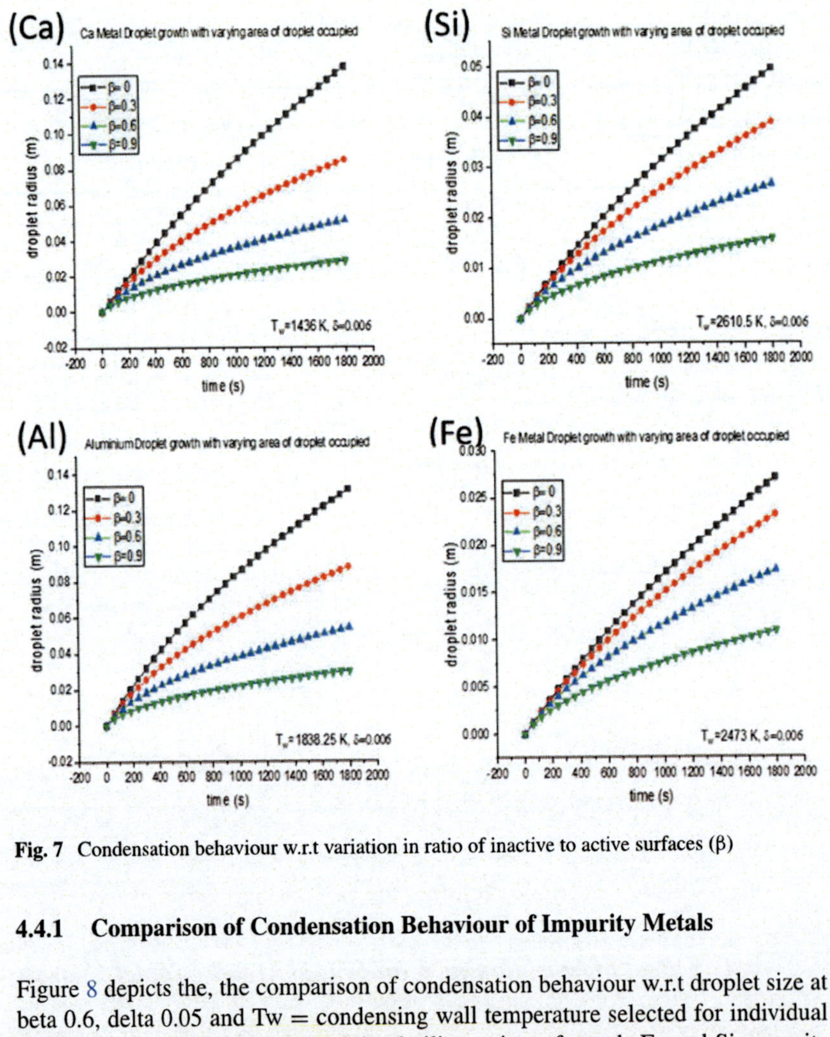

Fig. 7 Condensation behaviour w.r.t variation in ratio of inactive to active surfaces (β)

4.4.1 Comparison of Condensation Behaviour of Impurity Metals

Figure 8 depicts the, the comparison of condensation behaviour w.r.t droplet size at beta 0.6, delta 0.05 and Tw = condensing wall temperature selected for individual metal liquidus. It should be noted that boiling points of metals Fe and Si are quite higher than remaining impurities metal. For initial time the growth rate is approximately linear. Due to the similar boiling and condensation behaviour of metals Ca and Al, the metal vapours may be affecting the condensation and final purity of Magnesium. The separation of Ca and Al from Magnesium is considered a difficult task as the condensation behaviour. Further, If the β value and δ value are high, the rate of growth is low which is due to large constriction resistance involved in condensation process. At lower β value and δ value, the rate of growth is high and the thermal conductivity is also high.

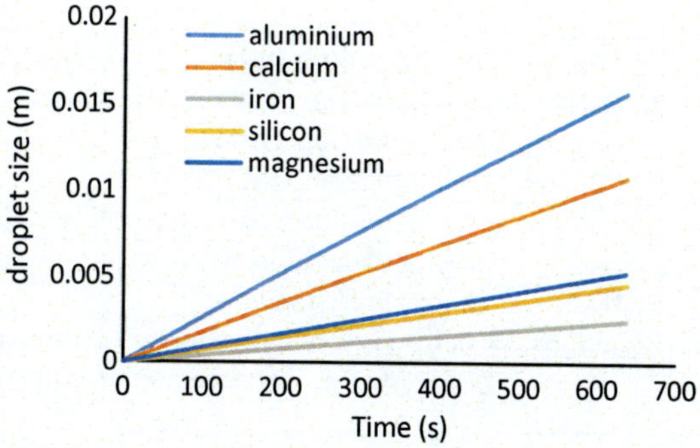

Fig. 8 Comparison of condensation behaviour of individual impurities

5 Conclusions

The relative volatilities estimated for Mg (Al, Fe, Si, Ca) systems using Thermo-Calc® software predict the feasibility of separation. The separation coefficients depict clearly that, Mg can be purified using distillation technique and separation of impurities is possible under controlled thermodynamic conditions. Along with thermo-physical properties, thermal gradient, surface thickness and ratio of active to inactive area on condensation surface have important control over metal condensation phenomena. Further, based on the liquidous range and boiling characteristics as estimated by relative volatility concept, the separation can be designed for required steps and selective boiling and condensation of specific components in the Mg-impurity matrix.

Acknowledgements Authors are thankful to CSIR-NML for supporting the presented studies, under in-house project support group (ipsg) project-OLP 0312 and winter internship program at CSIR-NML.

Nomenclature

α	thermal diffusivity of drop (m^2/s)
h	Heat transfer coefficient (W/m^2K)
ρ	density of liquid (kg/m^3)
k	thermal conductivity of liquid (W/(m·K)
T	Temperature (°C, K)
R	local radius of drop (m)
h_{lv}	latent heat of vaporization (J/kg)
δ	Thickness of promoter layer (m)
$\hat{\sigma}$	accommodation coefficient
σ	Surface tension (N/m);
q_d	surface heat transfer rate ($W/(m^2K)$

t	Time (s)
V	volume of drop (m^3)
T_s	Saturation Temperature (°C, K)
θ	the contact angle (rad or°)
T_w	Wall temperature (°C, K)
ΔT	($T_s-T_{i,}$/w) degree of subcooling (°C, K)
R_i	initial radius of drop (m)
R	Thermal resistance
r_{min}	Thermodynamically smallest radius (m)
R*	drop radius at interface and time t (m)
suffix	–
int	interface
cap	Capillary resistance
cond	conduction
sub	substrate
constr	constriction
l	liquid
v	vapour
s	solid

References

1. Gupta CK (2003) Chemical metallurgy, principles and practices. Willey-VCH Verlag GmbH & Co. KGaA, Weinheim
2. Moore JJ (1981) Chemical metallurgy (first). Butterworth and Co(Publishers) Limited, London, Boston
3. Rajput RK (2009) Heat and mass transfer. S.Chand and Company Ltd., New Delhi
4. Ray HS, Shridhar R, Abraham KP (2015) Extraction of non ferrous metals. Affiliated EAST-West Press Pvt. Ltd., New Delhi
5. Drápala J, Kuchar L (2015) Metallurgy of pure metals-study support, Issue- I,2015 (No. No. CZ.1.07/2.2.00/28.0304).
6. Revel G, Pastol J-L, Rouchaud J-C, Fromageau R (1978) Purification of materials by vacuum distillation. Metall Trans B 9B(December):665–672
7. Deng Y, Yang B, Li D, Xu B, Xiong H (2013) Purification of indium by vacuum distillation. In: Zhang C.J.L., Allanore A, Wang C, Yurko JA (eds) Materials processing fundamentals. Springer, Cham, pp 193–197
8. Wang Y, Tian Y, Qu T, Yang B, Dai Y, Sun Y (2014) Purification of magnesium by vacuum distillation and its analysis. Mater Sci For 788:52–57
9. Stephan K (1992) Heat transfer in condensation and boiling. Bergles AE, Mayinger F, (eds). Springer, Berlin, Heidelberg
10. Habashi F (1986) Principles of extractive metallurgy-pyrometallurgy. Gordon And Breach, Science Publishers, New York, London
11. Seryukov N, Kuzmin B, Chelishchev Y, Kuznetsov B (1969) General metallurgy, 2nd edn. MIR Publishers, Moscow
12. De S, Mondal S, Mohanasundaram M, Sau DC, Gupta RK, Kumar M, Paul KK (2016) Modeling heat transfer of the electrothermal reactor for magnesium production. Int J Therm Sci 102:274–284

13. Kumar K, Das SK, Kumar M (2020) Heat transfer modelling of dropwise condensation behaviour of magnesium vapours in the electrothermal production of magnesium. Can Metall Q 59(2):134–150
14. Krishna K, Sahay PK, Randhawa NS, Gharami K, Kumar M (2018) Flux refining of crude/sponge magnesium produced from the silicothermic process. In: Abhilash TS (ed) 22nd International Conference on Non-ferrous Minerals and Metals—2018. Ranchi, Corpoarte Monitor
15. Kumar K, Randhawa N, Gharami K, Sau DC, Kumar M (2016) Studies on dissolution kinetics of dolime in electrothermal magnesium slag. Russ J Non-Ferrous Metals 57(4):287
16. Khandekar S, Muralidhar K (2014) Dropwise condensation on inclined textured surfaces. In: Francis AK (ed) springerbriefs in applied sciences and technology-thermal engineering and applied science. Springer
17. Rose JW (2002) Dropwise condensation theory and experiment: a review. Proc Inst Mech Eng Part A J Power Energy 216(2):115–128
18. Sikarwar BS, Khanekar S, Muralidhar K (2014) Dropwise condensation of liquid metal vapour underneath a flat inclined substrate. In: 7th International conference on computational heat and mass transfer paper, p 135
19. Rose JW (1999) Condensation heat transfer. Heat Mass Transf 76(2):143–152
20. Rose JW (1967) On the mechanism of dropwise condensation. Int J Heat Mass Transf 10:755–762
21. Takaharu T, Tanaka H (1991) A theoretical study on the constriction resistance in dropwise condensation. Int J Heat Mass Transf 34(11):2779–2786
22. Bahrami HRT, Saffari H (2017) Mathematical modeling and numerical simulation of dropwise condensation on an inclined circular tube. J Aerosp Technol Manag 9(4):476–488
23. Kreyszig E (2013) Advanced engineering mathematics, 9th edn. Wiley
24. Bohdansky J (1968) Temperature dependence of surface tension for liquid metals temperature dependence of surface tension for liquid metals. 2982
25. Takaharu T, Hiroaki T, Shigenori T (1991) Experimental verification of constriction resistance theory in dropwise condensation heat transfer. Int J Heat Mass Transf 34(11):2787–2796
26. Busey RH (n.d.) Liquid and Gas. Heat Fus Vaporiz 75(1952):2–5
27. Sikarwar BS, Khandekar S, Muralidhar K (2013) Mathematical modelling of dropwise condensation on textured surfaces. Sadhana–Acad Proc Eng Sci 38(6):1135–1171
28. Valencia JJ, Quested PN (2008) Thermophysical properties. In: ASM handbook: casting, 15(Ref 24). pp 468–481

Microstructure Analysis of Alumina Effect on Compressive Strength of Iron Ore Pellets

Rakesh Prasad, Sadhna Bijrothiya, Manoj Narwariya, and Naresh Kumar

1 Introduction

Iron ore consumption for iron and steel production in the blast furnace and direct reduction unit turn into the exhaustive stage. Only inferior grade or iron ore is remains in which the aluminous nature of ore makes the process difficult to get extract iron. In this process of pelletization, making good quality pellets through additive mixing and heat hardening is turn in to tedious task [1]. The grade inferiority is increased by 2 wt % with respect to high-grade iron ore in India [2]. Earlier many researchers have done their work on the influence of alumina on pellets strength [3]. Pellet strength plays an important role to make a good quality of feed for the blast furnace. Pellet should be enough strong to transport one place to another and should be durable under heavy loading in the stack in the furnace. The high-temperature stability of pellet burden to maintain permeability for reduction makes its use worth full. To provide strength to pellet heat hardening is a very important process during which alumina has a significant role to make an intermediate phase in the microstructure of pellet structure [4–6]. Past research states that an increased amount of alumina decreases the strength of the pellets remarkable. Al_2O_3 content reduces the viscosity of the silicate phase of bonding between the particles [7].

Phase analysis of heat-hardened pellet microstructure interpretation supports the effect of alumina content over physical and physicochemical properties of pellets. In the present work, Gua Indian iron ore has been chosen for aluminous effect for physical properties such as cold compressive strength and tumbler index. The effect

R. Prasad (✉)
Hindustan College of Science and Technology, Mathura, India

S. Bijrothiya · M. Narwariya
IPS College of Technology and Management, Gwalior, India

N. Kumar
Fuel & Mineral Engineering Department, IIT Dhanbad, Jharkhand, India

R. Prasad et al. (eds.), *Advancement in Materials Processing Technology*,
Springer Proceedings in Materials 12,
https://doi.org/10.1007/978-981-16-3297-6_25

Table 1 Chemical composition of Gua Iron ore

Fe total	Al_2O_3	SiO_2	LoI
64.5	0.9	4.2	6.1

Table 2 Pellet mixtures composition (wt %)

S. No	Fe total	Al2O3	Bentonite	Coke	Lime stone
1	95	1	2	2	1
2	94	2	2	2	1
3	93	3	2	2	1
4	92	4	2	2	1

of alumina is expressed with the help of microstructure analysis and phase grains density of silicate, hematite, and magnetite, and pore phases present in the structure of the pellet.

2 Experimental

2.1 Raw Material

The sample has been taken from Gua Jharkhand India as a dump fine, the chemistry of which is shown in Table 1. The effect of alumina was investigated by preparing four samples of a mixture with a variable amount of alumina.

2.2 Mixtures

In order to make flux pellet, 2 wt% of limestone and coke was added to make basicity in the moderate range and coke to prevent more fuel consumption, bentonite as a binder. A variable amount of alumina had been introduced to prepare four samples with increasing order of alumina shown in Table 2.

2.3 Balling Process

Green ball formation was done in disc pelletizer for these parameters were taken in use as per Table 4 shown. Water spray ensures the moisture addition to provide proper shape during nucleation and coalescence of ore particles to get ore ball in disc pelletizer as shown in Fig. 1a.

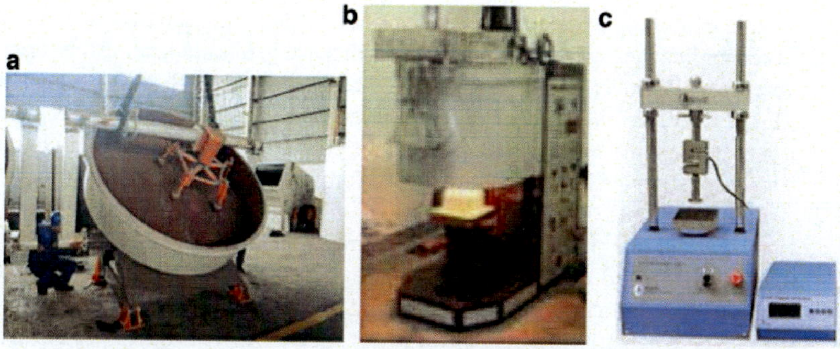

Fig. 1 **a** Disc pelletizer **b** Raising hearth furnace **c** CCS tester

2.4 *Heat Hardening and Testing*

Heat hardening of the green ball is a necessary and very important process of pellet making. In Table 3 heating cycle is shown, in which drying is done up to 350 °C and then preheating of the pellet was done in between 350 and 1000 °C. For the firing of the pellet of temperature was raised up to 1250 °C with some soaking time of 20 min in the raising hearth furnace as shown in Fig. 1b.

Fired pellets were cooled to go for compressive strength test with the machine capacity of 0–10 KN as shown in Fig. 1c in Table 4 balling machine and CCS tester description is given.

3 Microstructure Analysis

A pellet strength mainly depends upon the slag phase and crystal phase bonding between the grains. These bonding between the particles certain the pellet strength, bond stability in the presence of porous matrix are the very important term. For this microstructure analysis is an important aspect of strength interpretation of pellets.

Table 3 Heating process

Drying	Preheating	Firing (°C)
27–350 °C	350–1000 °C	1250 °C

Table 4 Experimental equipment's parameters

Equipment	Main parameters
Balling disc	Diameter 1 m; revolution speed 10 – 60 r/min; angle 10 – 45°
CCS testing machine	Range 0–10 kN

Pellet microscopic images can be confirmed by preparing samples such as specimen preparation, the first pellet was ground with the help of sandpaper in the series of The Scanning Electron Microscope was used to take picture of SEM images as shown in Fig. 1. Now identification of various phases such as hematite, magnetite, silicate and pore phase must be recognized by colour coding as hematite bit Grey colour used for magnetite, the light grey colour used for porosity, the black dark colour used for silicate bond [8]. This is the representation of different phases as per their colour notification on grayscale representation in some microstructure images. A tool for Image J was used to check the number of grains from each phase present in the unit area of that particular phase, volume density of that particular phase, a total area occupied by that phase. As shown in the figure, SEM images were taken for five types of pellets 1–4 as shown in Fig. 2.

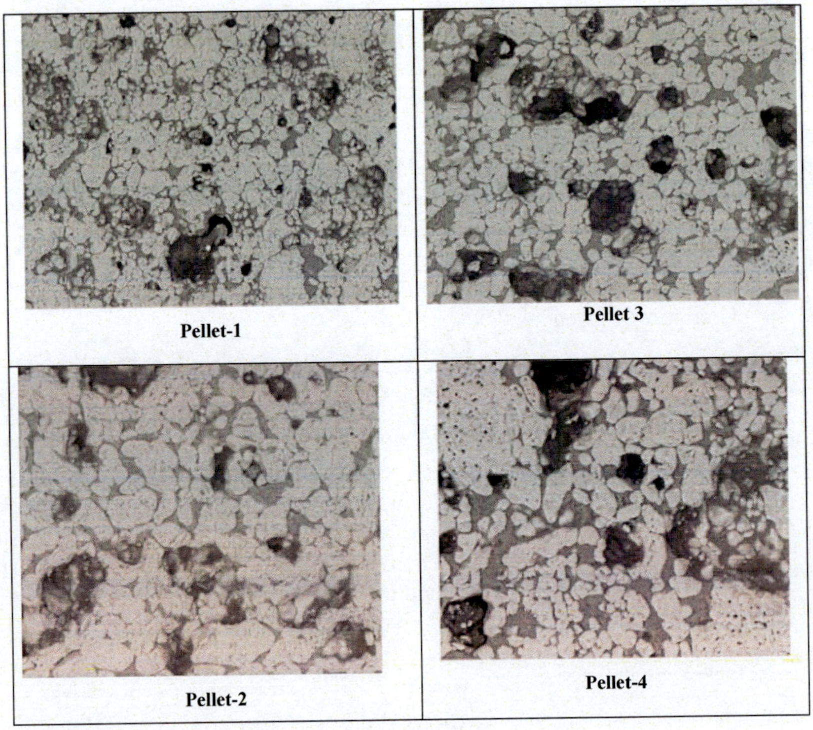

Fig. 2 SEM images of Pellets 1–4

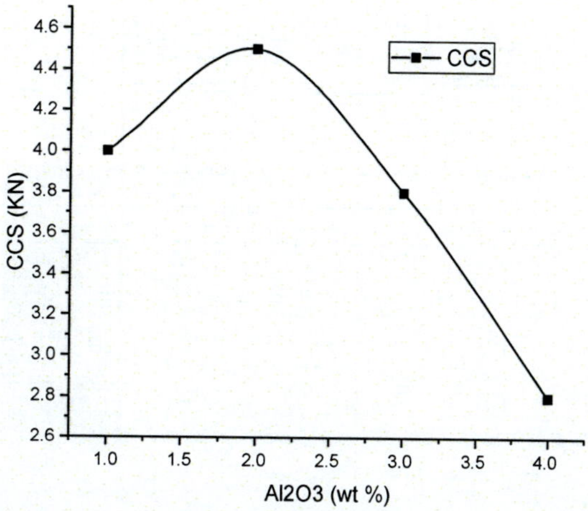

Fig. 3 Alumina content versus CCS

4 Result and Discussion

4.1 Cold Compressive Strength

The CCS variation with alumina content is shown in Fig. 3, in which effect on CCS due presence of alumina shows that with 1 wt % of alumina in pellet exhibit 4 KN CCS, as alumina content increases to 2 wt%, CCS value hiked to 4.5 KN. Afterward value of CCS decreases from 4.5 to 3.8 KN corresponding to 3wt% and 2.8 KN for 4 wt% of alumina in the pellets.

4.2 Grain Density

4.2.1 Alumina Content Versus Hematite Phase Grains Density

Hematite grain density is shown in Fig. 4. The grain density of 325 grains/mm^2 is shown for 1 wt% alumina pellets. As alumina content hiked to 2 wt% of pellets, the hematite grain density is increased to 442 grains/mm^2 which is the highest in the four samples. With increasing the alumina content 3–4 wt% grain density reduced to 233–136 grains/mm^2 corresponding to increased content in the pellets ore.

4.2.2 Alumina Content Versus Pore Phase Grains Density

The pore phase density for pellets with minimum alumina content of 1 wt% in the pellet has 80.2 grains/mm^2 with a further increment of alumina content 2 wt% pore

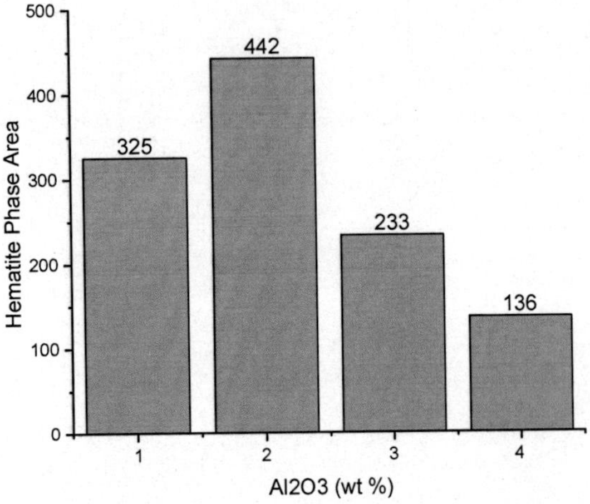

Fig. 4 Alumina content versus Hematite grains density

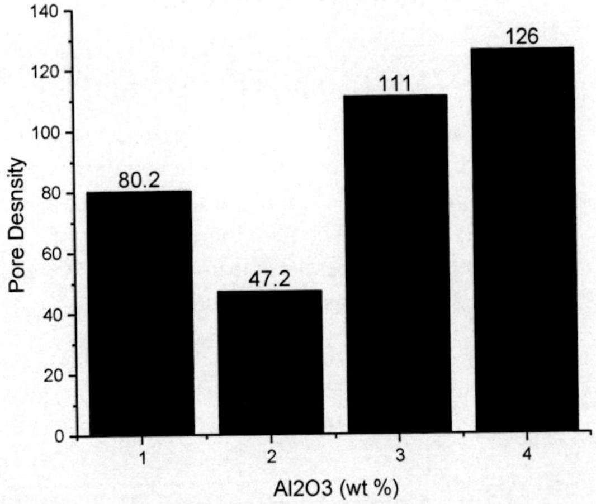

Fig. 5 Alumina content versus pore phase density

density decrease 47.2 grains/mm^2 which is the lowest in the four samples. It is remarkable that pore density increase with alumina content 3–4 wt % for showing pellets present in the matrix of slag phase-in the hematite grains network increased to 111 grains/mm^2 then 126 grains/mm^2 for highest alumina content is shown in Fig. 5.

4.2.3 Alumina Content Versus Silicate Phase Grains Density

The silicate melt phase plays the role of binder in the matrix of oxides grain density was 521 grains/mm^2 for the pellet with 1 wt% of alumina. As the alumina content reaches 2 wt% silicate melt phase also increases to 660% highest in the

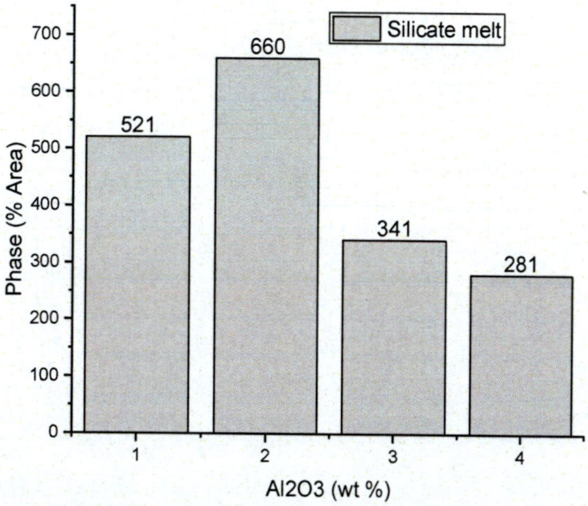

Fig. 6 Alumina content versus Silicate grains density

samples. Further increasing alumina content 3 wt% reduces the silicate melt to 341 grains/mm^2. When the highest alumina content 4 wt% used pellets show minimum silicate grains density 281 grains/mm^2 as shown in Fig. 6.

4.2.4 CCS Versus Phase Grains Density

Figure 7 shows that alumina content in the pellet silicate and hematite grains initially increases from 1 to 2 wt% and at this point, this is the highest value of grains density corresponding to maximum CCS. Further increment in the alumina these grains reduces and CCS decreases. The phase combination of increasing and decreasing

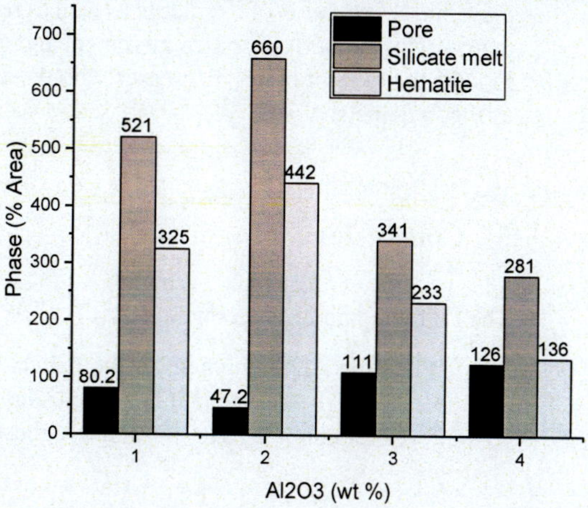

Fig. 7 CCS versus Phases grains density

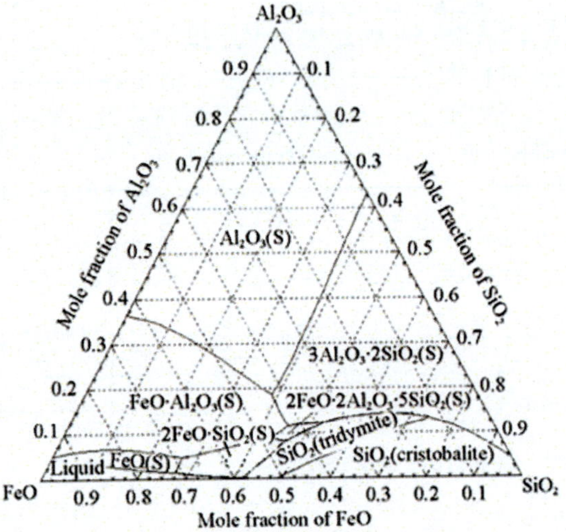

Fig. 8 Al2O3–SiO2–FeO ternary phase diagram [9]

particular phase is decided the strength of the pellets and microstructure orientation influenced the strength of the pellets. This can be observed a combination of hematite and silicate phase decide the strength of pellets and a combination of pore phase reduces the pellets bonding and finally weakened the structure.

This is known that alumina has acidic nature with silica; it tends to deteriorate the structure of iron fayalite. Eventually, this makes silica disable to hold magnetite and wustite phases in the main matrix. Some grains were liberated from a network of hematite that wakened the structure. However, fayalite supports the recrystallization of hematite and also improves crystal bonding between the particles. Therefore strength of pellets increases initially then it decreases as previous reasons. When an excess of alumina is added to the ore, the high melt of alumina is formed and prevents the oxidation of magnetite as well as recrystallization of hematite within the matrix. This supports pore size increment and uneven distribution of the pore phase. Finally, the CCS of pellets decreases with the additional amount of alumina content as Fig. 8 depicts.

5 Conclusions

The following points are concluded as

(1) When the Al2O3 content in the pellets was increased from 1 to 4 wt%, the compressive strength first increased slightly and then gradually decreased. For 2 wt% of alumina, this shows the highest CCS of 4.5 KN.

(2) Microscopic images show that reduced CCS of pellets are due to cracks development and holes formation, which was less at less alumina content then increases with alumina content in the size and interconnectivity.
(3) Alumina content of more than 2 wt% turns into, high melt and prevents the oxidation of magnetite as well as recrystallization of hematite within the matrix. This supports pore size increment and uneven distribution of pore phase subsequently reduces the strength of pellets.
(4) Up to 2 wt% of alumina content fayalite supports the recrystallization of hematite and also improves crystal bonding between the particles hence strength of pellets increases.

References

1. Meyer K (1980) The pelletizing iron ore. Springer Verlag, Berlin
2. Long F (2006) The effection on ironmaking raw materials and metallurgical process by high content of Al2O3 Iron Ore [Dissertation]. Wuhan University of Technology, Wuhan, p 1
3. Zhang YP, Fu JY, Jiang T, Yang YB (2002) The influence of the gangue contents on properties of pellet. Sinter Pellet 27(4):11
4. Ye KW (2007) Heavily push forward the pelletizing production in our country,. Sinter Pellet 32(5):1
5. Zhang J-L, Wang Z-Y, Xing X-D, Liu Z-J (2014) Effect of aluminum oxide on the compressive strength of pellets. Int J Miner Metall Mater 21(4):1
6. Sharma T, Gupta RC, Prakash B (1993) Effect of firing condition and ingredients on the swelling behaviour of iron ore pellets. ISIJ Int 33(4):446–453
7. Kawachi S, Kasama S (2011) Effect of micro-particles in iron ore on the granule growth and strength. ISIJ Int 51(7):1057–1064
8. Meyer M, Lagoeiro LE, Graça LM, Silva CJ (2016) Phase and microstructural characterization of iron ore pellet and their relation with cold crushing strength test. Miner Process Extr Metall Rev 37(5):295–304
9. Zhang YM (1998) Pellets production technology. Metallurgical Industry Press, Beijing, p 66